Themes with a Difference
228 New Activities for Young Children

Themes with a Difference
228 New Activities
for Young Children

Moira D. Green

DELMAR
CENGAGE Learning

Australia • Brazil • Japan • Korea • Mexico • Singapore • Spain • United Kingdom • United States

Themes with a Difference: 228 New Activities for Young Children
Moira D. Green

Publisher: William Brottmiller

Senior Editor: Jay Whitney

Assosiate Editor: Erin O'Connor Traylor

Production Editor: Marah Bellegarde

Editorial Assistant: Ellen Smith

For product information and technology assistance, contact us at
Cengage Learning Customer & Sales Support, 1-800-354-9706
For permission to use material from this text or product,
submit all requests online at **www.cengage.com/permissions**
Further permissions questions can be emailed to
permissionrequest@cengage.com

Library of Congress Control Number: 97-13490

ISBN-13: 978-0-7668-0009-0

ISBN-10: 0-7668-0009-1

Delmar
Executive Woods
5 Maxwell Drive
Clifton Park, NY 12065
USA

Cengage Learning is a leading provider of customized learning solutions with office locations around the globe, including Singapore, the United Kingdom, Australia, Mexico, Brazil, and Japan. Locate your local office at **international.cengage.com/region**

Cengage Learning products are represented in Canada by Nelson Education, Ltd.

To learn more about Delmar, visit **www.cengage.com/delmar**

Purchase any of our products at your local bookstore or at our preferred online store **www.ichapters.com**

Notice to the Reader
Publisher does not warrant or guarantee any of the products described herein or perform any independent analysis in connection with any of the product information contained herein. Publisher does not assume, and expressly disclaims, any obligation to obtain and include information other than that provided to it by the manufacturer. The reader is expressly warned to consider and adopt all safety precautions that might be indicated by the activities described herein and to avoid all potential hazards. By following the instructions contained herein, the reader willingly assumes all risks in connection with such instructions. The publisher makes no representations or warranties of any kind, including but not limited to, the warranties of fitness for particular purpose or merchantability, nor are any such representations implied with respect to the material set forth herein, and the publisher takes no responsibility with respect to such material. The publisher shall not be liable for any special, consequential, or exemplary damages resulting, in whole or part, from the readers' use of, or reliance upon, this material.

Printed in the United States of America
8 9 10 11 12   14 13 12 11 10

ED182

CONTENTS

PREFACE

This book uses a child-initiated, whole language approach to help you have fun exploring that world with children.

Notice that each unit begins with an "Attention Getter." The purpose of this is to introduce each unit to children in a way that grabs their attention, stimulates their interest, and creates excitement about the discoveries you will be making together. Because they focus the group's attention, I also like to use the "Attention Getter" at the beginning of every day or session to connect with the children before they explore the activity centers, to discuss the projects which are available for the day, and to sing songs or read books which reinforce the unit. Many of the activities in this book include demonstrations of the materials that can be incorporated in your Attention Getter time.

You'll also notice that with a few of the children's books in the literature lists, I suggest that you use colored pencils to shade in diverse skin colors, expecially when illustrations show large groups or crowds of people who are all white. Even in this day and age, some publishers of children's books are quite unaware of this issue, so I don't hesitate to alter books to reflect the population, and I urge you to do the same.

Each unit is arranged according to a WHY? WHAT? HOW? format, as in the following example:

The Exploding Triangle Trick
Science

WHY we are doing this experiment: to provide children with a dramatic way of observing how soap weakens surface tension; to develop self-esteem and a sense of autonomy through use of a one-person work station.

WHAT we will need:
 Shallow tub of water
 Water refills
 Tub for emptying used water
 Three 10 cm (4") straws
 One popsicle stick
 Small container of liquid soap
 "One person may be here" sign (see
 page 165)
 Newspaper

HOW we will do it: To prepare, spread several layers of newspaper underneath your activity area. Pour a shallow amount of water in the tub, set all other materials beside it, and pin up your "One person may be here" sign.

Activities selected are from across the curriculum: science, math, music, movement, language art, multicultural diversity, dramatic play, social studies, motor and cognitive development.

It is also important that you conduct each activity ahead of time by yourself before facilitating it with the children. This allows you to anticipate problems, to set up the activity more efficiently, and to make sure that your particular materials work as desired.

The implementation of proper safety precautions is always a primary concern when working with children. Some educators have recently expressed reservations about the use of glitter. I have included this material in my activities because in my years of teaching I have never experienced, or heard of, an injury occurring as a result of its use. However, I have also listed colored sand as an alternative to glitter, if you prefer to use it. Also, in regard to safety, make sure you try each activity yourself before facilitating it with the children. I've included thorough safety precautions throughout the book, but your particular materials, methods, or facilities may alter the equation, so it's important to have a "trial run" to spot any potential hazards.

Finally, there is a lot of discussion these days among early childhood educators about the best approach to teaching young children. High Scope, open-ended, child-directed—these are a few of the terms commonly used. I have implemented a whole language philosophy in this guide because in my teaching experience, students thrive on this approach. There is nothing like being with children who are so enthralled with a project that they spontaneously use all parts of language—listening, speaking, reading, and writing—in the thrill and excitement of their explorations and discoveries. I have also used a child-initiated approach throughout the book because my experience has been that children who choose what they would like to do, and the length of time they would like to do it, are empowered children. That said, I would like to end this preface with the observation that regardless of our particular teaching approaches, genuinely caring for, respecting, and having fun with our children is what matters most, and is the best gift we can give them. Enjoy experiencing these activities together, and have fun!!

Dedication

To my parents, Tom and Louise Green, with love.
To those fabulous realtors at the Stark East office in Madison, Wisconsin.
To my sister, Deirdre Green, for being who she is and then some.
And to H. S. for everything.

ACKNOWLEDGMENTS

To my Delmar editors, Jay Whitney and Erin O'Connor Traylor.

Special thanks to the reviewers of my manuscript who provided great ideas and advice:

Mary Henthorne
Western Wisconsin Technical College
La Crosse, Wisconsin

Judy Patchin
Black Hawk College
Genesco, Illinois

Dr. Audrey Marshall
Albany State College
Albany, Georgia

Ruth Steinbrunner
Central Virginia Community College
Lynchburg, Virginia

Etta Miller
Taylor University
Ft. Wayne, Indiana

Dr. M. Kay Stickle
Ball State University
Muncie, Indiana

Thank you to Bruce Sherwin and Linda Ayres-DeMasi at Publisher's Studio for their talent and hard work. Thanks also to Jennifer Campbell for her excellent copyediting. And special thanks to Hud Armstrong for his wonderful illustrations.

FAT-TASTIC

Attention Getter: Gather the following materials: margarine, cooking lard, a tub, a smooth ball, vegetable oil, a tub of warm soapy water, and paper towels. Cover the ball in vegetable oil and put it into the tub. When the children are gathered, pass the tub around and invite the children to try to pick it up. Is it easy? Why not? Show the children the bottle of vegetable oil, and tell them that when people bake, they often grease their pans with vegetable oil first. Using their experience trying to pick up the oily ball, ask the children why they think this is hard. Show your students the lard and margarine and say: "If we melted this lard and this margarine down, they would be oily liquids, like this vegetable oil. The margarine, the oil, and the lard are all different kinds of fat. Lard is animal fat. The margarine and the oil are vegetable fat, pressed out of safflower or sunflower seeds. Butter is the fat that comes from milk." Ask the children to guess what you are going to talk about for the next few days. Have them

wash their hands with the clean, soapy water, and then give them small smears of margarine and lard to explore with their hands and fingers. How does each one feel? Explain that oil is fat in a liquid state and that the margarine and lard are fat in a solid state.

Fat Floats
Science/Sensory

WHY we are doing this project: to provide children with a hands-on opportunity to discover that fat floats.

WHAT we will need:
> Dollops of margarine
> Dollops of lard
> Vegetable oil
> Medicine droppers
> Clear plastic cups
> Tubs
> Cold water
> Trays
> Small pitchers
> Small containers for vegetable oil
> Newspaper
> Warm, soapy water
> Paper towels

HOW we will do it: Spread many layers of newspaper underneath the work surface. Set the warm, soapy water and paper towels near the experiment activities, but not so near that the children might drop fat into the water. The soapy water and towels are for cleaning hands.

Put the clear plastic cups and medicine droppers on a tray with the small containers of vegetable oil. Place the small pitchers of water nearby. Have some empty tubs available in which experiments can be dumped when children want to begin new ones.

Pour cold water into the other tubs and set the dollops of lard and margarine nearby. Ask the children to predict what will happen when they put the fats in the water. Ask them what they see in the room that will help them find out. As your students conduct the experiment,

ask them what they notice about whether it is possible to sink the fat dollops. What do the vegetable oil drops do in the water? (Immediately float to the surface.) Is it possible to keep oil drops at the bottom of a cup filled with water? (Fat weighs less than water, so it floats.) What happens when one oil droplet meets another one in the water? Encourage verbalization of comments.

Oil (Liquid Fat) Repels Water
Science

WHY we are doing this experiment: to provide children with a hands-on method of discovering that fat repels water.

WHAT we will need:
> Wax paper
> Water
> Vegetable oil
> Small containers (for oil and water)
> Medicine droppers
> Toothpicks
> Food coloring
> Trays

HOW we will do it: To prepare, put a tray in front of each chair at an activity table. Put a sheet of wax paper and a toothpick on each tray. Add food coloring to the small containers of water and arrange these and the containers of vegetable oil so that all the children have access to them. Ask the children to predict what they will observe if they drop both colored water and oil onto the wax paper. Invite them to conduct this experiment. Do the two fluids mix? Does this result change when they use toothpicks to stir the oil and water together? Explain that fat molecules are drawn to other fat molecules, and water molecules are drawn to other water molecules. *Molecules* are tiny, tiny particles. *Repel* means to push away. Using this information, ask your students to hypothesize about why the oil and water cannot be mixed very easily.

More Proof That Oil and Water Do Not Mix (Part 1)

Science/Sensory

WHY we are doing this experiment: to provide children with another fun method of proving that water and oil do not mix.

WHAT we will need:

Small plastic containers with lids
Tall plastic bottles with caps
Small pitchers of water
Food coloring
Funnels
Vegetable oil
Tubs or sensory table
Newspaper

HOW we will do it: Spread many layers of newspaper underneath your work area. In the tubs or sensory table, provide separate, small pitchers of oil and of water. Add the containers, bottles, lids, caps, and funnels. It is nice to have tall, narrow bottles as well as wider, shorter containers because this allows children to see that even though the container shapes change the thickness of each layer, the water and oil still separate. This is especially interesting to observe in Part 2 of this project.

During an Attention Getter time, ask the children if they have ever seen their fathers or mothers make salad dressing. How did they make it? Ask the children what they see in the room that will help them make their own water and oil mixtures. Ask them to predict whether the water and oil will mix if their containers are shaken hard. Also ask: "What will happen if you shake your oil and water and the lid of your container has not been screwed on properly?" As the children conduct the experiment, ask them what they observe. What do they notice about whether the water and oil stay mixed?

More Proof That Oil and Water Do Not Mix (Part 2)

Science/Sensory

WHY we are doing this part of the experiment: to provide children with an opportunity to see how other materials settle, float, or sink when mixed with oil and water.

WHAT we will need:

Oil and water mixtures (from previous activity)
Small bowl of coarse ground pepper
Small bowl of crushed red pepper
Bowl of ketchup
Bowl of mustard
Mustard seeds
Small pitcher of red vinegar

HOW we will do it: Invite the children to add all or some of the above to the oil and water mixtures they made in the previous activity. Ask them to predict what will happen to the new materials when the children screw lids and caps on and shake their containers hard. Do the pepper flakes sink or float? What happens to the ketchup and mustard? How many distinct layers can the children count in their containers after everything settles? If you like, add small pitchers of water to which green or blue food coloring has been added.

Lard and Margarine (Solid Fats) Repel Water

Science/Language

WHY we are doing this experiment: to provide children with a hands-on method of discovering that solid fat also repels water; to develop speaking and listening skills; to develop reading skills.

WHAT we will need:
- Pats of margarine (on individual wax paper pieces)
- Spray bottles on "mist" setting
- Water
- Dark-colored plastic plates (and/or cookie sheets)
- Large plastic garbage bags
- Activity sign (provided below; photocopy and enlarge for your use)
- Newspaper
- Puppet

HOW we will do it: Spread newspaper out on the activity table and put the dishes on top. Dark-colored plates work best because the margarine is light-colored and the children can more easily observe where the grease adheres and where it is repelled. Set a plate in front of every chair at the activity table. Cut squares out of the plastic bags, which will cover half of the plates. Plastic garbage bags work best because they are heavy enough to lie down flat (especially new, unused bags). Do not put too much water in the spray bottles. Put one margarine pat beside each plate. Put all other materials out on the table so that all children will have access to them.

Make an activity sign based on the sample provided and pin it up near the activity table. During an Attention Getter time, interpret/read the activity sign. Ask your students to predict what they will see when they follow the sign's suggestions. Invite them to do so. Take out your puppet and use its personality to ask the children what they are doing. Can the margarine be spread on the wet part of the dishes? How about the dry part? Ask the children if they remember how oil and water reacted to each other in the previous experiment. Using this information, ask them to hypothesize about why margarine cannot be rubbed onto the wet part of the dishes and pans.

Science Experiment with Grease and Water:

Alter this sign to reflect your materials.

Put plastic on one half of a plate or cookie sheet. Spray water on it. Take away the plastic. Now try to smear margarine on both sides. What happens?

Making a Water Magnifier with the Help of Grease
Science

WHY we are doing this experiment: to enable children to observe how a greased, looped paper clip will hold a drop of water (the fat repels the water and exerts pressure, which holds the drop in place).

WHAT we will need:
>Metal paper clips
>Margarine
>Sensory table or tubs
>Small containers for margarine
>Small measuring cups or coffee scoopers
>Bowls of water
>Medicine or eyedroppers
>Small letters and shapes page from
>>Magnificent Magnifiers unit
>Small paper towel squares
>Warm soapy water

Preparation:
>Pencil

HOW we will do it: To prepare, unbend one paper clip for each child, and rebend it around a pencil to form a loop with a handle. Make sure the upper part of the loop actually touches the lower part where they cross, so that the circumference will hold a drop of water.

Put the metal loops, small scoopers, bowls of water, and eyedroppers in the sensory table or tubs.

Make photocopies of the tiny shape, letter, and number page from the Magnificent Magnifiers unit, and put these on an activity table near the other materials.

During an Attention Getter time, hold up one of the paper clip loops and, using your fingers, grease it thoroughly with margarine. Ask your students to predict what will happen to a water drop if it is dropped onto it. Ask the children if they see anything in the room that will help them conduct this experiment. Invite them to do so. What happens? Do water drops fall through the greased metal loops? (Because the grease repels the water, it pushes out on the water, and this pressure keeps the water drop wedged in the middle of the loop.)

Tip: Encourage the children to turn the metal loops so that the overlapping part is on top, and gently pour streams of water over them.

Invite your students to use their water-drop magnifiers to examine the sheets of tiny shapes and letters.

After they have explored the materials for a while, invite the children to wash their greasy loops in the warm soapy water, dry them with paper towels, and hold them under a stream of water. Will an ungreased metal loop hold a water drop?

What Dissolves Grease?
Science/Sensory

WHY we are doing this experiment: to enable children to observe the effects of cold water, warm water, and liquid soap on grease.

WHAT we will need:
>Tub of cold water
>Tub of warm water
>Tub for dishes
>Children's play dishes
>Pats of margarine
>Liquid soap
>Vegetable oil
>Small containers (for soap and oil)
>Medicine droppers
>Spoons
>Newspapers

HOW we will do it: Spread many layers of newspaper underneath your work surface. If you have a large sensory table, put all the tubs inside it. Place the margarine pats, vegetable oil, and some spoons beside the tub of play dishes, and the liquid soap, spoons, and medicine droppers beside the tubs of water.

During an Attention Getter time, ask the children if they have ever had grease on their hands. How did they get it off? Ask them if they've ever washed greasy dishes or seen their parents wash greasy dishes. Encourage observations, and then show the children the materials and encourage them to explore. As they do, ask: "What do you notice about whether the cold or warm water takes the grease off the dishes faster? What happens when the liquid soap is added?" Warm water and liquid soap make fat molecules move apart from each other. They *dissolve* the grease.

Developmental differences: Three- and young four-year-olds tend to become engrossed in the sensory aspect of this activity. Through their exploration, they will discover that oil or margarine stays on their hands in cold water and comes off in warm, soapy water. Older children tend to enjoy sensory exploration and also enjoy observing what happens to the oil and grease on the dishes.

Fat Makes Paper Transparent
Science

WHY we are doing this project: to enable children to observe that as fat saturates paper, the paper becomes transparent; to develop reading skills with an activity sign.

The facts of the matter: The fat of margarine soaks into the spaces between the fibers of paper. As it soaks in, or saturates, the paper becomes transparent because the fat transmits light, so the light is conducted directly through the paper.

WHAT we will need:
Construction paper
Regular white paper
Newspaper
Brown paper bags
Kitchen parchment paper
Margarine pats
Small containers for margarine
Small pictures, letters, and numbers page
from Magnificent Magnifiers unit
Tub of soapy water
Paper towels
Activity sign (provided below; photocopy
and enlarge for your use)

HOW we will do it: Kitchen parchment paper is available in supermarkets. To prepare, cut all the paper samples into squares measuring about 7.5 cm x 7.5 cm (3" x 3"). Put pats of margarine in the small containers, and make an

Science Experiment:
What does margarine do to paper?

Dab some margarine on the different papers.

What happens?

activity sign based on the sample in the text. Post the sign near the work surface. Make photocopies of the shapes, numbers, and letters page from the Magnificent Magnifiers unit. The purpose of these is to provide something children can examine through their transparent papers.

During an Attention Getter time, read/ interpret the activity sign together. Ask the children to predict what will happen to the papers when the fat is rubbed on them. Ask your students if they see anything in the room that will help them conduct this science experiment. As they follow the sign's suggestions, ask them to notice which paper the fat of the margarine makes most transparent and which paper becomes least transparent. Ask them to notice the thickness of the papers. Is there a connection? Explain that *saturate* means to soak in until there is no room for more, and ask your students if they can think of other ways to use this word. Then ask them to notice how the transparent spot spreads when they rub margarine into the paper.

How Much Fat Is in Meats?
Science/Language

WHY we are doing this experiment: to enable children to compare the fat content of a variety of meats; to develop speaking, listening, reading, and writing skills.

WHAT we will need:
Kitchen parchment squares (7.5 cm x 7.5 cm [3" x 3"])
Small pieces of bacon (cooked and uncooked)
Salami slices
Summer sausage slices
Hot dog slices
Small containers (for meat samples)
Butcher paper
Markers
Scotch tape
Chart headings (provided below; photocopy and enlarge for your use)

HOW we will do it: Arrange all materials on the table so that all children will have access to them. Prepare the headings for a language chart, and pin it up near the experimentation area.

During an Attention Getter time, explain to the children that they can press the paper squares down onto the meat slices and see how large the transparent spots are when the paper soaks up the fat from the meat. (Make sure you discuss the fact that the children should never eat raw meat.) Ask the children to predict which meat will cause the most and least fat to saturate the paper.

After the children have conducted the experiment, read/interpret the headings on the language chart together. Have the children tape a sample of a grease-spotted paper from each meat under the corresponding heading. Together, as a group, look at and compare the size of the grease spots and the degree of transparency.

Note: Some of this will be determined by how hard the paper was pressed onto the meat sample. For the purposes of the language chart, you may want to make your own grease spots on paper so that you can be sure the same amount of pressure was applied to all the meat samples and that the results are more accurate.

As the children discuss the fat contents of the different meat samples, record their observations on the chart or have them write their words on the chart themselves. Use alternate colors of marker for each observation so that individual sentences can be more easily identified. Use quotation marks and write the children's names after their comments. Ask the children to look at each meat sample. Can they tell just by looking which one has the most fat? Does this correspond with the results of the paper test?

The Food Pyramid
Social Studies /Small Group Project

WHY we are doing this: to make children aware of healthy eating habits; to develop observation skills; to develop cognition by classifying food groups.

WHAT we will need:
Food pyramid chart (provided on page 9; photocopy and enlarge for your use)
Home and food magazines

HOW we will do it: Pin the food pyramid chart up on the wall near the place where you will be gathering, and have your food magazines available.

During an Attention Getter time, point to each part of the food pyramid and, together, discuss the amounts and types of foods that should be eaten. Leaf through the home and food magazines as a group, and when you come to a food type, have the children say which food group it belongs to. Do the children think that that food is high in fat or low in fat? You may also choose to discuss high sugar content/low sugar content foods and where they fit into the food pyramid.

Fat Means Nonstick
Science/Math/Small Group Project

WHY we are doing this project: to enable children to observe that fat acts as a lubricant; to practice rational counting; to familiarize children with spoon and cup measurements.

WHAT we will need:
Three ripe bananas
Two eggs, well beaten
2 cups flour
¾ cup sugar
1 teaspoon salt
1 teaspoon baking soda
Large bowl
Muffin tins
Margarine
Wax paper squares
Food coloring
Butter

HOW we will do it: Before you begin, remind the children about not sneezing or coughing on food and dishes while cooking, and about

What to eat every day.

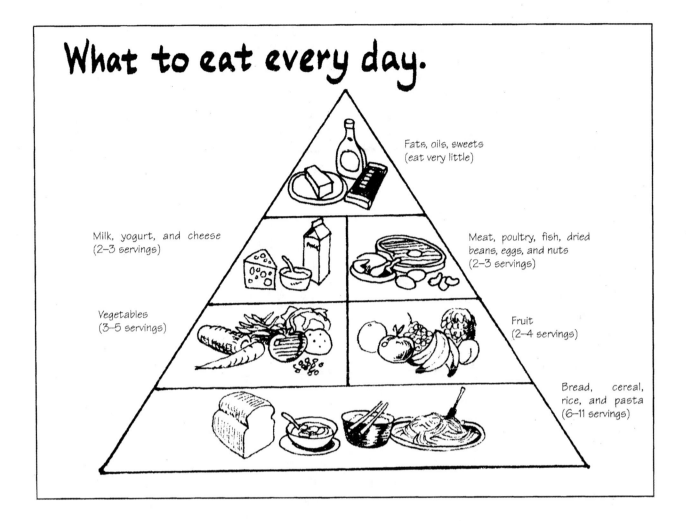

Fats, oils, sweets
(eat very little)

Milk, yogurt, and cheese
(2–3 servings)

Meat, poultry, fish, dried
beans, eggs, and nuts
(2–3 servings)

Vegetables
(3–5 servings)

Fruit
(2–4 servings)

Bread, cereal,
rice, and pasta
(6–11 servings)

keeping fingers out of mouths, eyes, noses, and ears. Pass the muffin tin around and invite each child to use a wax paper square and some margarine to thoroughly grease one muffin section in the tin. Leave one muffin section ungreased.

Preheat your oven to 350°F. In a large bowl, mix the bananas and eggs together. Let the children take turns stirring. Then let some children take turns measuring out the flour, sugar, salt, and baking soda and pouring them into the wet mixture, while other children continue to stir.

Make sure all the muffin sections (except one) are thoroughly greased, including around the rims and tops. When children fill the pans, batter tends to get everywhere. Ask your students to spoon the mixture into the muffin sections. Add some food coloring to the muffin batter in the ungreased section so that you will be able to easily distinguish it. Bake the muffins for forty minutes or until firm and brown.

(Baking time varies, depending on your oven.) Test for doneness by inserting a metal knife into one muffin. If it is clean when you remove it, the muffins are ready. If some sticky batter clings to the knife, more baking time is needed.

After the muffins have cooled, turn the pans upside down and have the children watch as you shake them. Which ones fall out easily? Which muffin stays in the pan? Remind the children of how they greased the pan and how only one muffin section was ungreased. Ask the children to hypothesize about why the muffin in the ungreased section is difficult to pry out. If you like, spread butter on your warm muffins. Enjoy!

Pressing Oil
Science/Gross Motor

WHY we are doing this project: to help children understand where vegetable oil comes from; to help them understand that this fat can be pressed out of nuts, olives, and seeds; to develop the large muscle group.

WHAT we will need:
> Black olives (pitted)
> Shelled peanuts
> Shelled walnuts
> Clear, plastic freezer storage bags (Ziploc)
> Rolling pins
> Safflower oil
> Safflower seeds (available at pet stores)
> Olive oil
> Jar of peanut butter with oil at the top (try an all-natural peanut butter)

HOW we will do it: Put the bottles of oils, safflower seeds, olives, peanuts, and peanut butter on an exhibit table for the children to examine. Position the source material in front of the corresponding container (e.g., a peanut in front of a jar of peanut butter, an olive in front of a can of olives). Put the nuts and olives in double bags and seal them. Put them on the table with the rolling pins.

During an Attention Getter time, have the group identify each kind of oil and the kind of seed, nut, or vegetable it came from. Be sure the children know that the glass bottles should be left exactly where they are and not be picked up or carried away. Show your students the materials on the activity table and invite them to pound the nuts or olives, and then to roll the rolling pins over the bags using as much pressure as they can. Leave the materials out for several days, but refrigerate the bags in between times. Eventually, the inside of the bags will become coated with the oil from the nuts.

Making Butter
Science

WHY we are doing this project: to help children understand that there is fat in milk; to enable them to observe that butter is milk fat; to help them understand what *coalesce* means.

WHAT we will need:
> Three cartons of whipping cream
> Three plastic containers with lids
> Heavy steel balls or marbles (six)
> Small plastic spoons
> Crackers

Optional:
> Timer
> Popsicle sticks

HOW we will do it: Put two marbles in each container. During an Attention Getter time, ask the children if they know what butter is made from. Show them your cartons of cream and, if you like, use the plastic spoons to give everyone a small taste of the cream. Do the children think it tastes like it has a lot of fat in it? Tell the children that you are going to make butter. Show them the containers with the marbles in them and let them take turns pouring one carton of cream into each container. Let the children know that once they start shaking the containers, everyone has to take turns to keep shaking them until the butter is made. Encourage the children to move the containers in a figure-eight motion. If necessary, use the timer to help the children with taking turns. (See the activity directly following this one for a song children can sing while they shake.)

At first, your students will hear the marbles moving, but after a while the fat will begin to coalesce and the cream will be so thick that it will not be possible to hear them. Explain to the children that the fat is *coalescing*. This means that the fat droplets are drawn to each other and form globules. As more and more fat droplets cling to the globules, the globules get bigger and bigger until the butter is completely separated from the buttermilk.

Rinse the butter off, salt it if you like, and let the children use the Popsicle sticks to spread small dabs of butter on crackers to eat.

This Is the Way We Shake the Cream

Music/Movement/Cognitive

WHY we are doing this activity: to develop cognition through memorization of words and actions; to help children feel comfortable using their singing voices.

WHAT we will need:
> Song: "THIS IS THE WAY WE SHAKE THE CREAM" (To the tune of "This Is the Way We Brush Our Teeth")

> "THIS IS THE WAY WE SHAKE THE CREAM"
> This is the way we shake the cream,
> shake the cream, shake the cream.
> This is the way we shake the cream,
> and make it into butter.

> This is the way we spread the butter,
> spread the butter, spread the butter,
> This is the way we spread the butter
> and eat it with the bread.

HOW we will do it: Pretend to be shaking a jar of cream for the first verse. For the first three lines of the second verse, pretend to spread butter, and for the last line, pretend to eat the delicious snack. After the children have learned the words, can they remember to sing it without saying the word "cream"? The word "butter"? Try it!

Butter Box

Math

WHY we are doing this project: to practice rational counting and subtraction; to develop self-esteem and a sense of autonomy through use of a one-person work station.

WHAT we will need:
> Bowl
> Ice cubes
> Tin box
> Butter squares
> Writing sheet (provided on page 12; photocopy and enlarge for your use)
> Pens
> Blank paper
> "One person may be here" sign (provided on page 165; photocopy and enlarge for your use)

HOW we will do it: To prepare, put the ice in the bowl, and cut butter squares to put in the tin. The number you put in the tin should be determined by how high your children are counting. Then put the tin in the bowl of ice, to keep the butter squares hard. Make several copies of the writing sheet and put these, blank paper, and pens beside the butter box.

During an Attention Getter time, discuss the "One person may be here" sign and read/interpret the writing sheets. Encourage the children to explore the materials and let them know that they may use the blank paper instead of the writing sheets if they'd like. Talk about the fact that the butter squares in the butter box are for counting, not eating.

How many butter squares are in the butter box?

I counted _____ butter squares.

3∴.
Take three butter squares away.

How many are left? _____

Crunkle and the Carrot Juice

Language/Anti-Bias/Social Studies

WHY we are doing this project: to develop speaking and listening skills; to help children become critical thinkers; to help children become aware of the hidden messages in advertising.

WHAT we will need:
Flannel board shapes (provided on pages 13–15; photocopy and enlarge for your use)
Flannel board
Felt
Scissors
Glue
Clear contact paper

Markers or crayons
"Two people may be here" sign (provided on page 175; photocopy and enlarge for your use)
Story: "CRUNKLE AND THE CARROT JUICE" (provided)

HOW we will do it: To prepare, color in the flannel board shapes in the text. Ask friends, family, or parents to help you with this if you are short on time. Glue the shapes onto felt, cover with clear contact paper, and cut them out. Set up your flannel board and tell the following story to the children.

"CRUNKLE AND THE CARROT JUICE"

As you probably know, rabbits love carrots. Well, there was once a rabbit by the name of Crunkle who made carrot juice and sold it to all the rabbits for miles around. [Put up standing Crunkle.] Now there

Enlarge these drawings on a photocopier.

13

Everyone likes a rabbit who's fat and plump and happy!

Fatten yourself up! with Crunkle's CARROT JUICE

Everyone likes a rabbit who's thin and skinny and sleek!

Skinny Rabbits Drink Crunkle's CARROT JUICE

NEW! IMPROVED!

Everyone likes a rabbit who's not too fat and not too thin

Get that "just right" feeling with Crunkle's Carrot Juice

Totally NEW Formula

was one thing that Crunkle wanted more than anything else in the world, and that was to be rich, rich, rich. But in order to be rich, rich, rich, he knew he'd have to sell a lot more carrot juice. [Put up bottle of Crunkle's carrot juice and picture of Crunkle thinking.]

"Hmmmm," he thought. "How can I make those rabbits want to buy more of my carrot juice?" And he thought and he thought and he pondered and cogitated until his thinker was sore from thinking but finally he had what he thought was a wonderful idea. He made a big poster and it said: "Everyone likes a rabbit who's fat and plump and happy!! Fatten yourself up with Crunkle's carrot juice!" And he put this poster up beside his carrot juice. [Put up the poster and bottles of carrot juice and standing Crunkle.] The other rabbits came by and looked at the poster.

"Step right up, folks!" Crunkle said. "Be the most popular rabbit in your warren! You can see what the sign says plain as day—everyone likes a rabbit who's plump and fat and happy and Crunkle's carrot juice will fatten you up in no time!! There it is in writing, so it must be true!! Buy one bottle of Crunkle's famous carrot juice and be amazed by the results!!" Well, the other rabbits looked at each other and talked and wondered and didn't quite know what to think. But they all knew they wanted the other rabbits to like them.

"Everyone likes a fat, plump, happy rabbit," they thought. "Maybe that Crunkle's carrot juice really will help me." And before you know it they were all buying a bottle. And when they were gone [take group of rabbits down] Crunkle looked at all his money and then he laughed and skipped all the way to the bank.

"Hee hee!" he laughed, "I'm going to be rich, rich, rich!" [Put up Crunkle on his way to the bank.]

Well, all the rabbits had a bottle of Crunkle's carrot juice now, so nobody bought any more for a while. Do you think Crunkle was happy about that? [Let children answer.]

"Hmmmm," he thought. "How can I make those rabbits want to buy my carrot juice again?" And just like before, he thought and he thought and he pondered and cogitated until his thinker was sore from thinking, but once again he had what he thought was a wonderful idea. He made a new poster, and this one said: "Everyone likes a rabbit who's thin and skinny and sleek! Skinny rabbits drink Crunkle's new improved carrot juice!" And he put *that* poster up beside his bottles of carrot juice. [Put up poster and bottles of carrot juice and standing Crunkle.]

Well, pretty soon the other rabbits came by again [put up group of rabbits] and Crunkle said, "Step right up folks—don't be shy!! The poster says it plain as day: 'Everyone likes a thin, skinny sleek rabbit and skinny rabbits drink Crunkle's carrot juice!' There it is in writing, so it must be true! Buy a bottle of Crunkle's new, improved carrot juice and be amazed by the new, improved results."

Well, the rabbits looked at the new poster and they talked and they wondered, and pretty soon they began to think: "I really want the other rabbits to like me. I wonder if Crunkle's carrot juice really will make me thin and skinny and sleek. Maybe I'll just buy one bottle of this new improved stuff and try it." And they did. And when they were gone [take group of rabbits off the board and put blank picture up] Crunkle looked at all of his money and laughed and skipped all the way to the bank again.

"Hee hee!" he laughed, "I'm going to be rich, rich, rich."

Well, do you think Crunkle was satisfied? Not him—he began to think of another dastardly plan. [Put up thinking Crunkle.] Just like before, he thought and he thought and he pondered and cogitated until his thinker was sore from thinking, but once again he had what he thought was another wonderful idea. He made a *third* big poster, and this one said: "Everyone likes a rabbit who's not too fat and not too thin! Get that "just right" feeling with Crunkle's totally new formula carrot juice!" And just like before, he put the poster up beside his bottles of carrot juice. [Put up third poster beside bottles of juice.]

Well, the other rabbits came by, and Crunkle said, "Step right up, folks! Get your totally new formula Crunkle's carrot juice. Nobody likes a rabbit who's too fat or too thin!! Crunkle's new formula carrot juice will give you that 'just right' feeling!"

But this time, just as the other rabbits were thinking that they should buy a bottle, one of the rabbits (her name was Cotton Top) said to them, [put up rabbit with blue bow] "You know, I've bought two bottles of Crunkle's carrot juice already and I'm exactly the same now as I was before. My bunny friends and family didn't care if I was fat, and they didn't care if I was thin, and they didn't care if I drank Crunkle's carrot juice, and they didn't care if I didn't drink it. I'm not buying any more of your stupid carrot juice, Crunkle. I'm going down to the lettuce patch to play tag with my friends." And off she hopped to do just that. [Take Cotton Top off the board.]

Well, all the other rabbits began to think about it, and realized that Cotton Top was right, because the same thing was true for them. None of their friends and family had cared if they were fat and none of them had cared if they were thin and nothing had changed whether they drank Crunkle's carrot juice or not. And the more they thought about that, the angrier they got at Crunkle.

And Crunkle said, "Now, now, folks, no need to get riled up. A rabbit's got to make a living, doesn't he?" But the rabbits were too angry at themselves and at Crunkle to listen to his excuses, so they made him pack up every last bottle of his Crunkle's carrot juice and take them all away. And because he couldn't sell a single bottle any more, he packed everything up and left town [put up ashamed Crunkle] and the last I heard of him he was trying to sell cabbage juice out by Bobtail Bluff. As for the rest of the rabbits, they went down to the lettuce patch too, and played tag until the sun went down and that's the end of that story.

Questions for discussion:

• Do you think it was true that everyone liked rabbits who were plump and fat? Thin and skinny? Think about the people you like. Why do you like them?

• When Crunkle's posters said the carrot juice was "new and improved" and a "totally new formula," do you think the carrot juice was really better or different? Why did Crunckle say it was new and improved?

• One of the posters said: "Everyone likes a rabbit who's thin and skinny," and Crunkle said, "There it is in writing so it must be true." Do you think he was right? If something is in writing, is it always true? Is everything on the

television true? (Lead into a discussion of things seen on television that are not true.)

• When Cotton Top said that her friends and family didn't care if she was fat or thin or if she drank Crunkle's carrot juice or not, the other rabbits realized the same thing was true for them. The story says they got mad at Crunkle and they got mad at themselves. Why do you think they were mad at themselves?

VOCABULARY:
Cogitate
Satisfied
Ponder
Dastardly
Amazed
Realize
Results
Warren
Sleek

After telling and talking over the story, discuss the "Two people may be here" sign. Leave the flannel board and the pieces out for the children so that they can retell the original story as well as make up new ones.

What Is Real?
Social Studies/Anti-Bias/Small Group Activity

WHY we are doing this activity: to make the children aware that all bodies have body fat and that it is a natural, normal part of all bodies; to make children aware of the function of body fat; to make children aware of body fat standards, especially for women, which are imposed by the media, and to encourage children to question them; to help children critically examine media images and messages in general; to develop observation skills.

Food for thought: In the 1950s, the average female model weighed 8 percent less than the average woman. Today, the average female model weighs 22 percent less than the average woman. Only one in every 80,000 women has the body type of the average model. Although there are many more sexualized images of women than of men in the media, men too are confronted with "perfect" body images.

WHAT we will need:
 Fashion magazines
 Photojournalism magazines
  (e.g., Sunday newspaper supplements,
  Life magazine)
 Barbie photograph (from a toy catalog)

HOW we will do it: Leaf through the magazines and cut out two kinds of photographs: those that feature posed, glamorized models and those that are slice-of-life photos of real people in action. Tape the former group of photographs and the Barbie picture on one side of a wall so that they will be clearly visible to the children. Do the same with the latter group of photos on another part of the wall.

Explain to the children that the fat in our bodies is stored energy. If we are in a very cold place, our body fat helps keep us warm. If for some reason we cannot get any food for a long time, our bodies use up our body fat to give us energy. Everyone has body fat. Body fat is a normal, natural part of everyone's body.

Together, talk about the photographs of the "real" people and compare them to the fashion models. How are they different? (Models are almost always posed; "real" people are often photographed in action. Female models are usually much thinner than real women and wear much more makeup. Female models are often wearing skimpy or tight clothes which reveal a lot of their bodies, clothes that are rarely seen on real women.) Compare the body of the Barbie doll to the pictures of real women. (Barbie doll: huge eyes, tiny nose, tiny waist, large bust, exaggeratedly long legs.) Ask the children if they know any real women whose bodies look like this. If you like, use a Ken doll to make the same comparison to real men.

Ask the children what they notice about the skin color of most of the models. Do most of the people in the world have white skin? Discuss these issues with the children. Remind them of the flannel board story about Crunkle. Did it matter to their friends and families how the bunnies looked and whether the bunnies were fat or thin? Ask: "Do you think your friends and your family love you for how you look, or for the special person that you are?" Discuss.

Oil Art
Art

WHY we are doing this project: to facilitate creative expression.

WHAT we will need:
 Large shallow pans
 Water
 Vegetable oil
 Tempera paint
 Medicine droppers
 Small containers (for colored oil)
 Construction paper
 Newspaper

HOW we will do it: To prepare, spread many layers of newspaper on the work surface. Put water in the shallow pans and set them on the activity table. Mix the oil and tempera paint well and put it in the small containers with medicine droppers. Find a surface where the pictures can dry and spread newspaper there as well.

When the children are ready to do the activity, invite them to use the medicine droppers to put drops of colored oil into the water. Encourage them to float pieces of paper on the top of the water for a few seconds and then to take them out. Let them dry on the surface you have designated for this purpose. The pictures will not dry completely because of the oil, so when you are ready to put them on the wall, hang them above the children's reach, but where they will still be visible.

Kitchen Play with Fat: Oil and Grease
Dramatic Play/Language

WHY we are doing this activity: to promote child-to-child interaction; to enable children to act out real-life situations through fantasy play; to develop vocabulary as well as speaking and listening skills.

WHAT we will need:
- Kitchen dramatic play area
- Empty margarine tubs
- Empty margarine boxes
- Empty lard containers (e.g., Crisco)
- Empty plastic vegetable oil bottles
- Measuring cups
- Measuring spoons
- Recipe books
- Aprons
- Muffin pans

HOW we will do it: Add the empty containers and cooking materials to a kitchen dramatic play area. Encourage the children to explore and play.

Birds Like Fat, Too!
Nature/Crafts

WHY we are doing this activity: to help children discover that fat is sticky and that seeds will adhere to it; to develop fine motor skills through crafts.

WHAT we will need:
- Fairly large pinecones
- Softened suet or lard
- Wild birdseed
- Shallow pans for seeds
- Popsicle sticks
- Twine

HOW we will do it: Put all materials out on the activity table. Encourage the children to use the Popsicle sticks to spread and poke suet or lard in between the scales of the pinecones and to then roll the pinecones in the pans of wild birdseed. Help your students wrap twine underneath the top scales of the cones so that you can hang them in trees. Try to hang them near a window so that you can watch your feathered visitors.

Literature

Symbol Key: *Multicultural
+Minimal diversity
No symbol: no diversity or no people

MacDonald, E. (1989). *Miss Poppy and the honey cake*. New York: Dial Books for Young Readers. (Butter is used to make the honey cake in this book. On the inside of the front and back covers, there is a recipe for honey cake which includes instructions about how to grease the pan.)

Seuss, D. (1984). *The butter battle book*. New York: Random House.

Thomas, M. (1972). *Free to be you and me*. (Cassette). Bertelsmann Music Group Co. New York: Arista Records. (This excellent tape includes songs which encourage children to examine advertising messages critically.)

Wake, S. (1990). *Butter*. Minneapolis, MN: Carolrhoda.

Extenders

Social Studies: Play the tape: *Free to Be You and Me*. Discuss with the children the issues this tape raises and sing the songs together. Carol Channing and Marlo Thomas are just two of the artists who make contributions. *Free to Be You and Me* is available on tape, CD, and video, and can be purchased at many children's book and toy stores.

Science: When the children experiment with how fat makes paper transparent, provide them with some warm, melted margarine to drop onto paper. Does the temperature affect the speed or degree of transparency of the paper? Explain that heat makes molecules move apart, so this makes the fat soak into or saturate the paper even faster.

Science: After you make the water magnifiers using grease, invite the children to try the same experiment with warm water or warm, soapy water. Why doesn't it work?

Language: Have children look through food magazines and cut out food pictures to make their own food pyramids. Take story dictation, support invented spelling, or spell or write down words to be copied, as appropriate.

Social Studies/Nutrition: After you introduce children to the food pyramid, bring in samples from each food section. Have the children taste each one and categorize which part of the pyramid it belongs to.

FABULOUS FUNGI

Attention Getter: Ahead of time, let some mold grow on a piece of bread. When you are ready to introduce this theme, collect two or three small potted plants, the mold, and a mushroom. When the children are gathered, put the plants to one side in front of the children, and the mold and mushroom to the other side in front of the children. Say: "Take a good look at the plants over here. And take a good look at the mold and the mushroom over here. How are they different?" Encourage observations and let the children gently touch each item.

Facilitate a discussion about fungi and how they are different from plants. Are they the same color? Ask the children to guess what you will be talking about and working with over the next few days. If you like, tell this joke: "Why did everyone invite the mushroom to their party? Because he was a fun-guy!"

The facts of the matter: Plants have roots and leaves and grow from seeds. Plants use sunlight and water to make their own food. Mold and mushrooms are fungi. A fungus cannot make its own food. It must live off something else, dead or alive. (For example, mold grows on bread and mushrooms can grow from dead wood.) A fungus grows from a spore. Spores are so tiny that generally, they cannot be seen.

Growing Molds (Part 1)
Science

WHY we are doing this experiment: to enable children to observe the growth of molds; to develop reading skills.

WHAT we will need:
Small cups (paper or plastic) (six per child)
Small plastic trays (or Styrofoam meat trays)
Masking tape or sticky labels
Cans of tomato soup
Small bowls
Spoons
Bread crumbs
Grated cheese
Soil
Corn syrup
Tub of soapy water
Paper towels
Plastic wrap
Labels (provided below; photocopy and enlarge for your use)
Glue sticks

HOW we will do it: To prepare, you may want to cut the sides of the cups down so that the molds will be more visible when they grow. If you do this, try to cut them down evenly so that your plastic wrap will lay evenly over the tops of the cups when you cover the prepared experiments.

You should have enough cups so that each child can have six. Put the soup in the bowls, and then add the spoons. In other bowls put some bread crumbs, soil, corn syrup, and cheese. Find a dark, warm spot where you can leave the experiments for several days. Display each label next to the material it identifies. Have the tub of soapy water and paper towels available for clean-up.

During an Attention Getter time, ask the children if they think it is possible to grow molds. (A mold is a fungus, because it grows from spores. Spores are too small to be seen, but they are everywhere.) Show the children the materials on the table and ask them how they think they could use the materials to find out if mold spores will grow in tomato soup.

Prepare an experiment yourself by having the children help you put a few spoonfuls of soup into six cups. Show the children the bowl of cheese and then the cheese label. Ask them to read/interpret the label. Ask one of the children to add a small amount of shredded cheese to one of your cups of tomato soup and to use a glue stick to glue the cheese label onto the cup. Ask: "Why is the label useful?" Ask another child to add some bread crumbs to the next cup of soup, and repeat the same process with the appropriate label. After the children help you add corn syrup to the third cup and soil to the fourth, prepare a fifth by scraping your finger on a dirty place on the floor and then sticking your finger into the soup. Your sixth and last cup should contain plain tomato soup. Label all cups, place them on a tray, then write your name on masking tape or a sticky label to put on your tray. Carry the tray to the warm, dark place, and then use a large piece of plastic wrap to cover all the cups. Ask the children to predict whether mold will grow in any of the cups and, if so, which ones.

Invite the children to prepare their own mold experiments. You may need to help the children carry their prepared trays to the designated spot. Have the children use the tub of soapy water and paper towels to wash their hands after touching the floor with their fingers.

Growing Molds (Part 2)
Science/Language

WHY we are doing this project: to facilitate recording the results of a scientific experiment; to develop all components of language arts: reading, writing, speaking, and listening.

WHAT we will need:
Blank paper
Patterns for observation books
(provided on page 24; photocopy and enlarge for your use)
Pens
Crayons
Markers
Preparation:
Stapler

HOW we will do it: Set up a writing center near the mold-growing area, where you can leave all materials out for the duration of the experiment. Use the stapler and the patterns provided to make observation books for each child. Use a photocopier to reduce the labels on page 22 and glue each label in its corresponding column on your observation sheet. Also, fold and staple sheets of paper to make blank observation books. Put these with the pens, crayons, and markers on the activity table.

During an Attention Getter time, a few days after you prepare the experiments in Part 1 of this activity, hold up an observation book and, together, read/interpret it. In front of the children, check your own experiment, and in a blank or prepared book, draw a picture of each sample of soup and write a few words about each one. You may want to do this on the first day that molds are visible. After you make your observations, remind the children that their observation books will look different, because everyone has a different way of doing things.

Designate a group time every day to check the experiments. Let the children use the writing area during the course of the session or day's activities. As the children work on their books, take story dictation, support invented

Mold Growing
Science Experiment

Observation Book

Photocopy as many of these as you need.

Day_____ ☀ ☾

BREAD CRUMBS	GRATED CHEESE	DIRT	CORN SYRUP	GERMS FROM THE FLOOR	PLAIN TOMATO SOUP
Inside the cup I see:	Inside the cup I see:	Inside the cup I see:	Inside the cup I see:	Inside the cup I see:	Inside the cup I see:

spelling, spell words or write down words to be copied, according to their needs.

Let the molds grow for as long as you like. Before the children take their observation books home, invite a few children every day to display and talk about their work to the group.

Growing on a Coconut: The Same or Different?

Science

WHY we are doing this activity: to see if new mold cultures (which did not grow on the tomato soup) will grow on a coconut.

WHAT we will need:
Coconut
Magnifying glasses

HOW we will do it: During an Attention Getter time, break open the coconut. It is best to break it into two halves, but I have yet to discover any ideal way of accomplishing this. Tell the children you want to find out what kinds of molds grow on a coconut. First you are going to let it sit out in the air for an hour, in the hope that some mold spores in the air will land on the coconut and start to grow. Ask the children to predict whether they will or not. After an hour or so, close the coconut back up and put it somewhere safe.

If you work with older preschoolers or kindergartners whom you know will not disturb the coconut for several days, put it on your activity table. If you work with younger children, keep the coconut out of their reach for several days, and then put it on the science table once the molds begin to grow. Put the magnifying glasses next to the coconut. How many different molds are there? Are any molds growing which did not grow in the tomato soup? Discuss!

Eating Mold

Science/Sensory/Math/Language

WHY we are doing this project: to help children understand that some molds are edible; to provide a sensory experience through the sense of taste; to familiarize children with cup and spoon measurements; to make children aware of hygienic handling of food; to develop speaking and listening skills through the use of a puppet.

WHAT we will need:
Roquefort cheese (one block)
Crackers
Blender
Celery sticks
Carrot sticks
Puppet
Small dish
Magnifying glass
Small paper plates
Napkins
Dip:
½ cup crumbled roquefort cheese
1 cup yogurt
½ cup mayonnaise
2 tablespoons vinegar
1½ teaspoons honey

HOW we will do it: Before you begin, have all the children wash their hands. Have roquefort slices, carrot and celery sticks, and crackers available on the table for sampling.

Use your puppet to help you facilitate a discussion about why children must not sneeze or cough on the food.

Ask the children: "Did you know that there are certain kinds of molds which are good to eat?" Open up the package of roquefort cheese and encourage the children to examine it. Break a piece of it off and put it into the small dish. Put this dish on the table with the food samples, next to the magnifying glass.

Together, make the dip by putting all dip ingredients into the blender. Let the children take turns filling the cup, ½ cup, tablespoon, and teaspoon with the appropriate ingredients, as well as taking turns to activate and stop the blender. Stop the blender several times and stir the dip as you mix it. Pour it into a bowl, and set it on the table with a spoon in it. Encourage the children to try the "mold dip"! (Show them how to use the spoon to take some dip and put it onto their plates.) Roquefort cheese is not a favorite food of young children and many may choose not to try it, but in this activity you are giving them that choice and demonstrating that mold can be a food. Take your puppet out again and use its personality to ask the children how the mold dip tastes and how they made it.

Fungus Farm
Science

WHY we are doing this project: to allow children to observe how mushrooms grow.

WHAT we will need:
    Mushroom-growing kit
    Magnifying glasses
    Sign (provided below; photocopy and
        enlarge for your use)
    Mushroom

The facts of the matter: The rounded top of a mushroom is called the cap; the gills are the fins underneath. The fleshy part which covers the gills is called the veil. When the veil breaks and the gills darken, the mushroom is ready to release spores. A typical mushroom makes about one million spores per minute, for a period of several days.

HOW we will do it: Mushroom-growing kits consist of mushroom spores in a soil-growing base and are available in nurseries and garden centers in the spring. They range in size and price from $16 for a 9 lb. kit to $22 for a 16 lb. kit, though prices probably vary more widely

Science Experiment:
Fungus Farm

WEIGHT: 9 LBS.
White Button Mushroom KIT

How do mushrooms grow?

depending on where you live. Mushrooms usually begin to appear anywhere from seven to fourteen days and continue growing over a period of five to six weeks.

If you work in a school that is on a tight budget, ask parents to donate a dollar or so for the mushroom kit. It is well worth the effort because children learn so much from seeing the mushrooms emerge and grow, and derive satisfaction from tending them and taking them home.

Read the instructions on the kit carefully, and follow them. Make sure you put the kit on a very low table so that the children can easily see the surface of the kit, where the mushrooms grow. Make a "Please look but do not touch!" sign and post it near the kit. If you work with very young preschoolers, you may want to put a masking tape line down, to indicate where the children should stand, to make sure that the growth of the mushrooms is not disturbed.

Introduce this activity during an Attention Getter time by telling the children that mold is one kind of fungus and that you are holding another kind behind your back. Invite your students to close their eyes and ask them to pass the fungus around. When the children open their eyes, encourage them to explore the *gills* underneath. Explain that that is where the mushrooms spores are, although they are too tiny to see. Also explain that like mold, a mushroom cannot use sunlight and water to make its own food, the way green plants can.

Show the children the mushroom-growing kit, and discuss the "Please look but do not touch!" sign. Also discuss the many different kinds of mushrooms. Do the children think *all* mushrooms are good to eat? Explain to them that there are many kinds of poisonous mushrooms. Ask: "If you were walking in the forest and you found some mushrooms growing, do you think it would be safe to eat them?" Discuss the fact that it is **never** okay to eat mushrooms you find outside. The only mushrooms you can be sure are safe to eat are the ones you buy in the store.

Show the children the kit and ask them what they think it might be. Introduce the information that the kit will allow them to grow mushrooms that are safe to eat. Use the instructions on the kit to discuss with your students what will need to be done to nurture the mushrooms. Designate a specific time every day to check the mushrooms as a group. Show the children the magnifying glasses and ask them what they could use them for. (To examine the emerging mushrooms.) You can pick the mushrooms at any stage of their development. It is also extremely interesting to pull mushrooms out with the root still attached, so that the children can examine this. As the mushrooms develop, give each child some mushrooms to take home. See the Mushroom Baskets activity for a way to carry them.

A Mushroom Has Many Parts
Science

WHY we are doing this activity: to enable children to observe all parts of a growing mushroom; to expand vocabulary; to develop reading skills.

WHAT we will need:
  Whole mushroom (from mushroom farm in previous activity)
  Diagram (drawn from sample provided on page 28)
  Magnifying glasses
  Trays
  Toothpick
  Fork

HOW we will do it: In this activity, you will need to dig a whole mushroom out of the mushroom farm. Dig very carefully so that the rhizomorph and mycelium (see diagram) remain intact. These root-like strands are very delicate. To begin, use the fork to gently scrape away the soil around the base of the mushroom. Depending on the size of the mushroom, you may find other tools to help you. Use the toothpick to scrape away as much soil as possible, and then gently pull the mushroom up. If you find that your mushrooms are easily uprooted, pull up several so that more than one child at a time can examine whole mushrooms. Set a mushroom and magnifying glass on each tray.

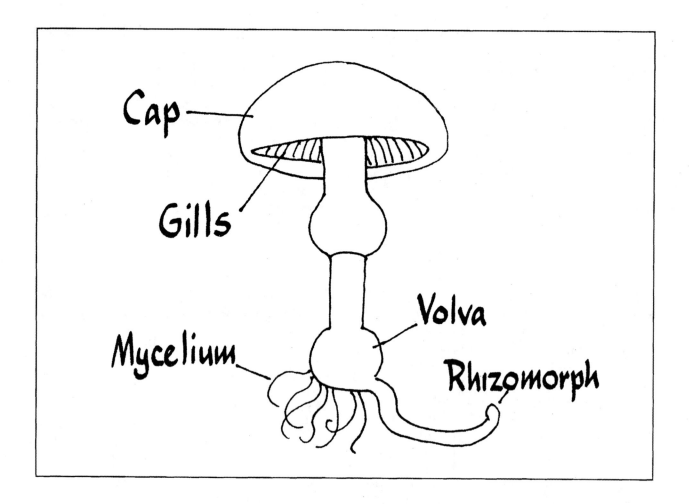

Cap

Gills

Mycelium

Volva

Rhizomorph

Using the illustration provided, make a large diagram of the different parts of a mushroom and post it on the wall near the activity trays.

During an Attention Getter time, go over the chart with the children. Pronounce the names of each mushroom part. Let the children know that if they would like to see these different parts of an actual mushroom, the materials they need are on the trays. Before they approach them, discuss how delicate and fragile the mushroom mycelia are. What will happen if they are rough with the mushrooms?

What Is in a Mushroom?
Science

WHY we are doing this experiment: to enable children to discover that mushrooms have a high water content.

WHAT we will need:
    Thick mushroom slices
    Paper towels
Demonstration:
    Tomato slice
    Paper towel

HOW we will do it: To prepare, cut the paper towels into quarters. Set the mushroom slices and paper towels on an activity table.

During an Attention Getter time, ask the children to predict what will happen if you squeeze the tomato slice between two pieces of paper towel. Conduct this experiment and show the children the paper towel pieces afterward. What came out of the tomato? If the children say, "juice," ask them what they think is in juice. If necessary, explain that the juice is mostly water. There is a lot of water in a tomato. Show the children the mushroom slices and paper towels on the activity table and ask them to predict whether a mushroom has a lot of water in it. What will happen if they squeeze the mushroom slices between two sheets of paper towel? As your students conduct the experiment, ask them about their findings.

More Proof That Mushrooms Contain Water
Science/Sensory/Math

WHY we are doing this activity: to enable children to observe and measure how much water is cooked out of a cup of mushrooms.

WHAT we will need:
> 4 cups sliced mushrooms
> Strainer
> Tablespoon and teaspoon
> Measuring cups
> Colander
> Funnel
> Containers
> Sensory table
> "One person may be here" sign
>> (provided on page 165; enlarge and photocopy for your use)
> One 2-cup measuring cup (plastic)

HOW we will do it: Ahead of time, cook two cups of mushrooms and save the water. The higher the heat, the more the water will evaporate. If you cook the mushrooms on a stove top, do so over medium heat. If you microwave the mushrooms, cover the bowl with plastic wrap so that evaporated water will condense on the underside of the wrap.

To prepare the activity, make an activity sign that says: "How much water in 2 cups of mushrooms?" Draw two cups above the words and mushrooms above that word. Post this and your "One person may be here" sign on the wall near the area where the children will be working. Next to the activity sign, place a two-cup measuring cup filled with fresh, sliced, raw mushrooms, so that your students can compare the size of these with the cooked mushrooms.

In the sensory table, place the cooked mushrooms (in their water), the containers, measuring cups, strainer, colander, and funnel. These straining materials provide children with a variety of ways of separating the mushrooms from their water; the spoons and cups provide two different methods of measuring that water.

During an Attention Getter time, ask the children if they remember the What Is in a Mushroom experiment. What did they find out about what is in a mushroom? Show the children the cup of uncooked, sliced mushrooms, and let them know that you cooked the same amount of mushrooms. Let the children know that if they would like to see what happened to the mushrooms you cooked, the materials are in the sensory table. Read both signs together, as a group, discuss, and invite your students to explore. How are the cooked mushrooms different from the raw ones?

Option: If you'd like, plug in an electric frying pan at the children's eye level and isolate the area with chairs and/or masking tape lines on the floor so that the children can observe from a distance. Sauté sliced mushrooms in the pan and, as the children watch, talk about what comes out of the mushrooms as they cook. Have a snack together by eating the mushrooms on toast.

Mushroom Baskets
Art

WHY we are doing this project: to facilitate creative expression; to develop fine motor skills; to develop self-esteem by enabling children to make a functional basket.

WHAT we will need:
> Small brown paper lunch bags
> Cardboard
> Glue
> Small containers for glue
> Glue brushes
> Paper
> Crayons
> Pens

Markers
Collage materials (e.g., bits of yarn, bits
   of paper, stickers)
Construction paper
"Fresh Mushrooms" label (provided
   below; photocopy and enlarge
   for your use)
Newspaper
Preparation:
   Stapler

HOW we will do it: Cut each paper bag down
to about 7.5 cm (3") high. Cut out cardboard
strips about 2.5 cm x 23 cm (1" x 9"). Make several copies of the "Fresh Mushrooms" label in
the text, and cut out blank pieces of paper the
same size so that the children can make their
own labels if they wish. Staple one end of each
cardboard strip to each paper bag to make a
handle. If you work with older children, you
may decide to help them do this for themselves.

Put all materials on the activity table.
Ahead of time, make a sample mushroom basket by gluing collage materials onto it or drawing on it. Pick a few mushrooms from the fungus farm and put them in your basket. During
an Attention Getter time, show the children
your basket and ask them what they see in the
room that would help them make their own.
When each basket is completed, save it until it
is that child's turn to take mushrooms home.

Broiled Fungus Sandwich
Snack/Sensory

WHY we are doing this project: to develop
self-esteem by enabling children to make their
own snacks; to provide a sensory experience.

WHAT we will need:
   Mushrooms
   Two sticks of softened margarine
   Salt
   Bread
   Broiler
   Plastic knives
   Plastic forks
   Bowls
   Paper plates
   Napkins
   Toaster

HOW we will do it: Although this snack tastes wonderful, it does not look too appealing, so you may want to make small amounts. Then if the children like it, you can make more. Take your margarine out of the refrigerator in advance so that it is very soft when the children use it.

Place a bowl and plastic knife in front of each chair, and divide the mushrooms up into the bowls so that each child has roughly the same amount. During an Attention Getter time, ask the children: "Have you ever had a fungus sandwich? Don't you think a nice, juicy fungus sandwich would be delicious?" Have some fun with the children when they respond to this, and then show them your bowl of mushrooms. Begin to slice the mushrooms, and as you do, explain that you are all going to make a broiled mushroom sandwich, but in order to make it, the mushrooms have to be cut up. Show your students the bowls of mushrooms on the table and encourage them to use the plastic knives to cut them. When their mushrooms are cut up into small pieces, give them plastic forks to use to mash squares of margarine into the mushroom pieces. There should be enough margarine in each bowl to hold the mushroom pieces together.

Toast your bread pieces, and then help the children spread their mushroom mixture onto a piece of toast. Broil until the mushrooms are cooked (a few minutes) and slightly burned on top, then salt. Enjoy!

Mushroom Print
Science

WHY we are doing this experiment: to allow children to see spores.

WHAT we will need:
   Large mushrooms, one for each child
   Paper
   Pens

HOW we will do it: In order for this experiment to work, the *gills* of the mushrooms must be exposed. Prepare each mushroom by cutting away any flesh from the underside which covers the gills. Cut the stems flush with the gills. Cut out paper squares that are large enough for each mushroom to sit on. Put all materials on the activity table. Find a place where you can leave the mushroom prints for several days.

During an Attention Getter time, take a mushroom, turn it upside down, and pass it around. Encourage the children to examine and touch the gills. Say: "In between those brown gills is where the spores of the mushroom are. The spores fall to the ground, and if the spores have what they need, they'll grow. Today we are going to do an experiment to prove that there are spores in the mushroom." Write your name on a paper square, put the paper on the table you have designated for the experiment, and put the mushroom on top of the paper. There should be paper under all parts of the mushroom. Invite the children to do the same, and ask them to predict what they will see when they check the paper in two days.

After two days, pick the mushrooms up. What do you see underneath? Notice how the mushroom prints get darker and darker as more days pass. Ask the children to hypothesize about why this is so.

Mushroom Center

Science/Multicultural

WHY we are doing this activity: to enable children to observe that fungi grow in many different shapes and sizes; to familiarize children with foods used in Japanese cuisine.

WHAT we will need:
  Samples of several different kinds of mushrooms (enoki [look like antennae], shiitake, portabella, oyster, canary [yellow], and wild)
  Index cards
  Marker
  Magnifying glasses

HOW we will do it: Specialty mushrooms are not cheap, so just buy two or three of each kind. If necessary, ask parents to donate a dollar or two toward the purchase of the mushrooms. Having a variety of fungi for children to directly examine is a very valuable part of this unit and well worth doing. The above list consists of mushrooms available in my local food co-op—your list may be different. Provide as many differently shaped and sized mushrooms as possible.

Put the mushrooms and magnifying glasses out on an activity table. Use the index cards and marker to print signs naming each mushroom type.

During an Attention Getter time, go over the names of each kind of mushroom with the children. Before they examine the fungi, talk about whether it is okay to eat or break the mushrooms. (They are too expensive to replace. If you work with very young children, you may want to put the mushrooms under clear plastic containers.) Show the children the enoki and shiitake mushrooms, and explain that they are used quite often in Japanese cooking. Ask the children if any of them have been to a Japanese restaurant. Did they eat any enoki or shiitake mushrooms?

Mushroom Writing Center

Language/Art

WHY we are doing this activity: to develop all components of language arts: reading, writing, speaking, and listening skills; to facilitate artistic expression.

WHAT we will need:
  Butcher paper
  Paper
  Markers
  Crayons
  Pens
  Scissors
  Glue sticks
Preparation:
  Adult scissors

HOW we will do it: Make a chart of the mushrooms you provided for the previous activity. If you'd like, ask the children to help you. Draw the shape of each mushroom, color in the appropriate shades, and print the name of each fungus. Put the chart up on the wall by the writing center table, and put all other materials on the table. Photocopy each mushroom drawing and name on the chart, put copies on the table, and have plenty of blank paper available. Cut blank paper into large mushroom shapes.

I have found that my writing centers tend to have more appeal for the children if I am able to set them up in a cozy corner. Dictate, spell, or write out words to be copied, and support invented spelling or scribbling. Children will utilize speaking and listening skills as they work together at the writing center, but you may also want to ask the children to read their work to you, or read their work to them. Some children may create projects that have nothing to do with mushrooms; they will still be developing valuable language arts skills.

Developmental differences: Three- and young four-year-olds enjoy scribbling with pens and crayons and dictating words. Older children enjoy drawing pictures and writing or dictating words.

Mushroom Match

Math

WHY we are doing this project: to develop cognition through a matching game.

WHAT we will need:

Cardboard panel (from large appliance box)

Yellow contact paper

Clear contact paper

Poster board

Mushroom patterns (provided below; photocopy and enlarge for your use)

Markers

Double-sided tape

"One person may be here" sign (provided on page 165; enlarge and photocopy for your use)

HOW we will do it: To prepare, cover the panel in yellow contact paper. Use the patterns provided to cut the mushroom shapes out of poster board. Trace a matching outline of each shape onto the board. If you work with kindergartners, make this matching game more challenging by drawing several shapes which are the same, then decorate them differently (e.g., dotted, striped, or small circles).

Cover the board with clear contact paper. Do the same to both sides of each mushroom shape. Put strips of double-sided tape on the board, the shapes, or both. Post the "One person may be here" sign above the board. Discuss this sign during an Attention Getter time. Children instinctively know what to do with these materials.

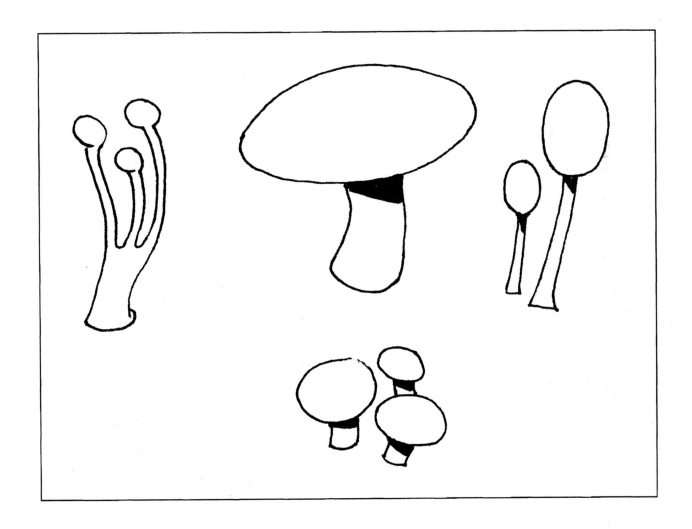

Mushroom Collage
Art

WHY we are doing this activity: to facilitate creative expression, to develop fine motor skills.

WHAT we will need:
    Construction paper
    Glue
    Small containers for glue
    Glue brushes
    Collage materials (e.g., scraps of fabric, paper scraps, beads, glitter, small macaroni pieces)
Preparation:
    Adult scissors

HOW we will do it: Cut the construction paper into large mushroom shapes and put all materials out on the activity table. Encourage the children to decorate their paper mushrooms in whatever way they would like. When the collages dry, put them up on your walls.

Toady Sittin' on a Big Toadstool
Music/Movement/Gross Motor/Small Group Activity

WHY we are doing this activity: to help children enjoy using their singing voices; to develop cognition through memorization of words; to develop the large muscle group.

WHAT we will need:
    Beanbag, ottoman, or masking tape
    Gym mats
    Song: "TOADY SITTIN' ON A BIG TOADSTOOL" (to the tune of "Froggy Went A-Courting")

"TOADY SITTIN' ON A BIG TOADSTOOL"
Toady sittin' on a big toadstool,
uh huh, uh huh,
thinkin' 'bout jumpin' in the nice cold
    pool,
uh huh, uh huh,
_____ came and tipped her off,
Toady fell in with a great big plop,
uh huh, uh huh, uh huh.

HOW we will do it: To prepare, put the beanbag or ottoman in the middle of the room, and put gym mats around it. If these materials are not available, make a circle on the floor with masking tape to simulate a toadstool.

Gather the children in the circle, with the "toadstool" in the middle. Sing the song, with a child's name in the blank space. Begin with one child sitting on the toadstool. The next child, who is named in the song, then jumps up onto the toadstool and gently tips the original child off. Before you play the game, talk about ways of tipping. Is it okay to tip hard and roughly? What might happen if students tip other children off that way? Have fun!

Mushrooms in the Kitchen
Dramatic Play/Language

WHY we are doing this activity: to provide opportunities for child-to-child interaction; to enable children to work through emotions and real-life experiences through fantasy play; to develop language skills.

WHAT we will need:
    Play kitchen furniture
    Sliced mushrooms
    Wok set or frying pans

HOW we will do it: Put a wok kit and some sliced mushrooms in your dramatic play area. Toy wok kits can be ordered from school supply catalogs, but you can also give the children small, regular frying pans to use, or toy ones.

Literature

Symbol Key: *Multicultural
              +Minimal diversity
              No symbol: no diversity or no people

Discuss the photographs in the following books and read selected parts.

Challand, H. J. (1986). *Plants without Seeds.* Chicago: Children's Press.

Johnson, S. A. (1982). *Mushrooms.* Minneapolis, MN: A Lerner Natural Science Book.

Selsam, M. E. (1986). *Mushrooms.* New York: William Morrow.

Extenders

Science: Experiment with conditions which prevent mold from growing. In separate experiments, try growing mold without air, moisture, light, or a food-growing base. Try growing it in the refrigerator and in a warm place. Which conditions make a difference to the growth of mold?

Math: Post a big piece of butcher paper on the wall beside your mushroom-growing kit. As a group project, make a mark for every mushroom you pick. Make tally marks in groups of five. When the kit stops producing mushrooms, count all the marks. How many mushrooms did you get from the kit?

Math: After the children have had a chance to examine the mushroom varieties displayed in your mushroom center, use the collection for a one-person work station and invite the children to sort and count the different kinds of mushrooms.

SHADOW SHENANIGANS

For this unit, it is best to have several flashlights. You can usually find inexpensive ones in variety stores. Make sure that you have plenty of extra batteries on hand. Before you begin the unit, talk with the children about the proper use and care of flashlights. Open one up and show them how the batteries fit inside. If you think you will need one, also have a timer available to help your children with taking turns.

Attention Getter: Find many different objects that you think will have interesting shadows (some suggestions: a crate, a lattice-backed chair, a clothes rack, a tall container). If you need to, use a flashlight, table lamp, or standing lamp to cause these objects to cast shadows. Turn the lights off until the children are gathered. When you are all together, tell the children that you are going to turn a light on, and that when you do, they will see something that they cannot see with the light off. What they see will give them a hint about what you will be talking about and working with over the next few days. Turn the light on or walk by the objects with the flashlight and ask the children what they notice. Which one is their favorite shadow? (Crates and lattice-backed chairs cast very pretty shadows.)

Making Light and Shadow Experimentation Boxes
Art/Gross Motor

WHY we are doing this project: to generate excitement about the unit by having the children prepare the materials that will be used in it; to facilitate creative expression; to exercise large muscles by bending and stretching.

WHAT we will need:
Refrigerator boxes
Newspaper
Paints (many different colors)
Large brushes
Poster board
Marker
Preparation:
Exacto knife

HOW we will do it: To prepare, decide how many refrigerator boxes you will need according to the number of children you are working with and the available space in the room, and pick them up from furniture stores that sell large appliances. It is a good idea to call first. One or two boxes for eight children has been plenty for me, but you will also have to take into account how much space you have. Using the Exacto knife, carve a door out of each box by cutting a top, side, and bottom and then folding the door flap back.

Spread out newspaper on the floor, and put the refrigerator boxes on the paper. Set the paint and brushes on small tables nearby.

At Attention Getter time, ask the children what makes a shadow. Discuss or explain that you need light and dark to make shadows. If you need to, pick up an object and shine a flash-

light on it and ask the children what they notice—if necessary, point out the light (flashlight) and the dark (shadow). Show the children the refrigerator boxes and invite them to look inside. Explain that you need the boxes for science experiments to explore light and shadow. Ask the children what you should call the boxes and write down what they suggest. Use their suggestions to make signs and attach them to the outside of the boxes. (Some suggestions I have received: "Experiment place,""Dark box"!) Then ask: "What could we do with the paint?" If necessary, invite the children to cover the refrigerator boxes with paint.

Light in Light and Light in Dark (Part 1)
Science

WHY we are doing this project: to enable children to discover the difference between what kind of patterns light produces in light, and what kind of patterns light produces in dark; to help children understand through hands-on experimentation that light is diffused in light and concentrated in darkness; to reinforce for young children the concepts of *inside* and *outside*.

WHAT we will need:
Colanders
Flashlights
Clear wall space
Experimentation boxes (created in previous activity)
Activity signs (provided on page 38; photocopy and enlarge for your use)

HOW we will do it: Place the colanders and flashlights on a small table near the experimentation box. Hang the activity signs nearby. Encourage the children to follow the activity signs' suggestions. When they shine their lights through the colanders in the room, why don't all the holes make light spots? What happens

Shine your flashlight
through the colander
<u>outside</u> the box.

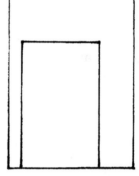

Shine your flashlight
through the colander
<u>inside</u> the box.

when they do the same thing in the experimentation box? What is the difference between the experimentation box and the room? Talk about how the light from the flashlight spills into the light in the room, but when the flashlight is used in the dark box, it is the only light, so it is forced through all the colander's holes.

Diffusion and Concentration of Light (Part 2)
Science

WHY we are doing this experiment: to build on the information learned in the previous activity and to facilitate another way of proving that light can be diffused or concentrated.

WHAT we will need:
    Cardboard
    Negative shape patterns (format provided on page 40; photocopy and enlarge for your use)
    Flashlights
Preparation:
    Exacto knife
    Scissors

HOW we will do it: To prepare, use the negative shape patterns provided to make cardboard shadow cards with these differently shaped holes in the middle. Obviously, the more interesting the shape, the more time consuming it is to cut out, so depending on the time you have available, select the simpler or more complex shapes. Make sure that you make all three sizes of whichever shape you pick. After you have made the shadow cards, lay them on the table with the flashlights. Show the children the cards and flashlights, and the experimentation box in which they can make light shapes with the materials. Ask the children to predict what difference there will be

between the three sizes of holes when light is shone through the bigger hole. As the children conduct the experiment and emerge from the experimentation box, discuss the results. Ask them to hypothesize about the difference between the smallest light shape and the biggest. Why is the smaller shape so much clearer? Explain that when light is shone through the bigger hole, the light is spread out over more space. The smaller hole has less space for the light to shine through, so that light shape is stronger and clearer.

Light and Distance
Science

WHY we are doing this experiment: to help children understand the connection between distance and the size and appearance of shadow.

WHAT we will need:
    Flashlights
    Colanders
    Fly swatters (with designs cut in the plastic)
    Construction paper
    Experimentation box
    Activity sign (format provided on page 41; photocopy and enlarge for your use)

HOW we will do it: To prepare, put a small table near the experimentation box, and set the colanders, fly swatters, and flashlights on it. Use construction paper to make the the activity sign and post it nearby. During an Attention Getter time, have the children sit together. Hold up an object, and ask them to notice whether or not it looks the same size as you hold it close to them as when you take it farther away. As you take the object further away from them, ask: "How does it look now? What about now?" Tell the children that they will have a chance to do a science experiment in the experimentation

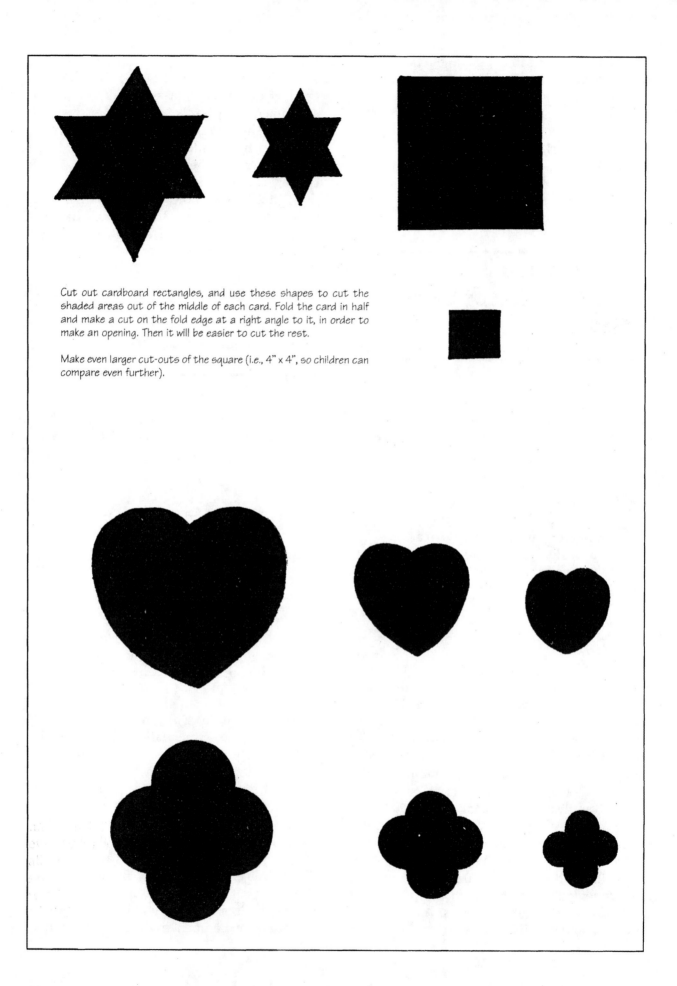

Cut out cardboard rectangles, and use these shapes to cut the shaded areas out of the middle of each card. Fold the card in half and make a cut on the fold edge at a right angle to it, in order to make an opening. Then it will be easier to cut the rest.

Make even larger cut-outs of the square (i.e., 4" x 4", so children can compare even further).

When you go into the experimentation box, hold your flashlight close to the colander, and then hold it far away.

box that will show them another way of proving that things look smaller when they are farther away, and bigger when they are close. Encourage the children to follow the suggestions of the activity sign and ask them to predict what they will see when they do. After your students conduct the experiment, discuss their findings. What is the difference between the light shadows from the holes in the colander when the colander is held close to the flashlight and the light shadows that are seen when it is held far away? Invite the children to do the same experiment with the fly swatters. Later, show them how to make different patterns by swirling and moving the colanders and fly swatters in front of the flashlight or cooperating with another person and overlapping them.

Tip: Save the refrigerator boxes to use again for the Space Unit.

Color and Shadow
Science

WHY we are doing this experiment: to allow children to observe how color fills in shadow.

WHAT we will need:
    Two clip-on lamps
    Acetate folder covers (red and blue)
    Rubber bands
    Packaging tape
    Large piece of white paper

HOW we will do it: Arrange your lamps near a wall so that the lamps face each other about two feet apart. Cut out circles from the acetate, several inches larger in circumference than the lamp openings. This allows you a margin of acetate to wrap around the edges of the lamps. Use rubber bands and packaging tape to cover one of the lamp openings with the red circle and one with the blue. Pin the white paper up behind the lamps.

During an Attention Getter time, show the children the lamps and point out the colored plastic covering on each one. Invite them to switch the lamps on and ask them to predict what they will see. After the children switch the lamps on and examine the effect, discuss their observations. What do they notice about the colors cast by each lamp? (The red light floods the shadow of the blue lamp; blue light floods the shadow of the red lamp.) If you like, put different colors of acetate on the lamps throughout the unit.

Shadow Story
Anti-Bias/Language

WHY we are telling this story: to stimulate interest in the art, drama, and science activities to follow; to present different gender roles than stereotypical ones; to demonstrate shadow puppet theater; to develop speaking and listening skills; to expand vocabulary.

It can take a little extra time to set up the light and puppet theater exactly right, but children do appreciate seeing drama presented in this unusual way and then trying out shadow puppet theater themselves.

WHAT we will need:
Small table or puppet theater
Film projector, strong lamp, or
    standing light
Shadow puppet outlines (provided on
    page 43; photocopy and enlarge for
    your use)
Cardboard
Popsicle sticks
Double-sided tape
Glue
Scissors
Clear wall space

Optional:
Clear contact paper
Story:
"TOPSY AND THE CARROT PATCH"
    (provided)

HOW we will do it: In this activity, you will need to make the characters in the shadow puppet story, and then tell the story to the children. This is intended to interest them in using the shadow puppets to retell the original story and to make up new ones, to make their own characters with the materials, and to experiment with the shapes and their shadows in the experimentation box. These activities are described below. To prepare the story, cut the character shapes out of cardboard. Glue a Popsicle stick onto all except the tree, basket, bone, acorn, and dead mouse.

The amount of time you put into making these will be determined by how often and how long you plan on using them. If you will be using them with many groups of children over a period of years, you will probably want to cover them with clear contact paper to protect them and make them sturdier. Otherwise, do not bother with contact.

Put several strips of double-sided tape on the non-Popsicle-stick side, close to the bottom of the shape. The purpose of the tape is so that each shape can be attached to the edge of the table when your hands are full of other characters. For the small bunny shape, also stick double-sided tape along its back because this character lays down to sleep. Cover one whole side of the bone, acorn, and dead mouse shapes because you will be sticking these onto the mouths of the animal characters as if they are carrying them.

If you are using a small table for a puppet theater, turn it over on its side. Arrange the table and projector or light source so that the cardboard shapes cast clear shadows on the wall. A film projector is really ideal—it is *very* effective for shadow theater. Also, think about where the children will sit to watch. It should be a place where they are facing the wall and can easily see the shadows—perhaps on either side of the table or puppet theater. A vocabulary list is provided following the story. Before you begin the story, stick the tree onto the edge of the table or puppet theater, in the middle.

BONE

TREE

DEAD MOUSE

ACORN

BASKET

Use photocopier to enlarge shapes.

"TOPSY AND THE CARROT PATCH"

Once upon a time there was a little bunny named Topsy whose very favorite food in all the world was carrots. She didn't much like having to dig them up out of the carrot patch, however, because it was very hard work. [Hold up small bunny shape at one end of the table.] She was always making excuses to get out of digging for carrots in the carrot patch. Now, one evening, this little bunny's father [hold up big rabbit shape] was making carrot stew for dinner when he realized he had no carrots left.

"Topsy," he said, "Run down to the carrot patch under the tree and dig up some carrots for me."

"Oh, Daddy," Topsy said, "I'm so sleepy. Can't someone else do it?"

"No, Topsy," Daddy Bunny said. "Everyone else in the family is out digging up turnips or lettuce or parsnips for the big bunny feast tomorrow, and anyway, it's your turn to go to the carrot patch. Here's a basket [stick up basket shape] and do hurry because the stew is cooking and I need the carrots right away."

So very halfheartedly Topsy hopped and loped down to the carrot patch.

Now when she got there, what do you think she should have done? [Let children answer.] But instead that lazy little bunny lay down on the ground in the middle of the carrot patch, under a big old tree. [Stick basket on edge of table, under tree, and stick bunny shape lying down, also under same side of tree. Place it at an angle so that the shadow of the Popsicle stick is not visible.]

"Yaaawwwwn. I know Daddy needs the carrots right away, but I won't have the energy to carry a big basket of heavy carrots all the way home unless I take a little nap first." And pretty soon that naughty little cottontail was fast asleep. Do you think she made the right decision? [Let children answer.]

Well, while Topsy was sound asleep, Ms. Squirrel came along. [Move Squirrel shape along edge, toward tree, from opposite direction.] She had some acorns in the pouch of her cheek, and she was looking for a good place to hide them for the winter.

[In squeaky squirrel voice:] "I think I'll just bury my acorns under this tree. It looks like a good spot where no one would think to look." So she dug and she dug and she dug [move squirrel shape as if it is digging] and she buried those acorns until not a speck of them could be seen and then she ran away to look for more acorns.

A few minutes later Black Cat came along. [Move cat shape along edge, from same direction.] She had a half-eaten, dead mouse in her mouth. Her tummy was too full for her to finish eating it just then, but she was afraid that if she didn't hide it, someone else would steal her delectable treat.

[In drawling cat voice:] "Mmmeeeooowww," she said. "This looks like a good place to bury my tasty morsel." And so she did. [Move cat shape as if it is digging.]

Do you think it was a tasty morsel you would like to eat? [Let children answer. Then move cat shape away.]

Well, a few minutes later Doggy came along [move dog shape along from same direction from which cat and squirrel came] and he had a big, stinky bone he wanted to bury so that it would get even stinkier.

[In deep, dopey dog voice:] "Duhh, this is just the place to bury my bone. Duhh, hope I remember where I hid it." And so he dug and dug and dug [move dog shape as if it is digging] and he buried that nasty, stinky bone and then he ambled away.

Well, right about then Topsy woke up. [Make yawning, stretching noises.] Suddenly she realized she must have been asleep for longer than she planned. [Make bunny shape jump around in agitation until bunny is on other side of tree where animals buried their things.] The sun was almost down and it was quite dark.

"Oh my, oh my, oh my," she said, "Tails and whiskers—Daddy's waiting for the carrots and I have been asleep for hours!" And she was in such a panic, and so afraid to go home empty-pawed, that she started digging right where she was. [Move bunny shape as if it is digging each time.] Guess what she found? [Hold up bone. Let children answer.] Yep, the first thing she found was a big, old stinky bone and because she couldn't really see what it was, she threw it into her basket. Guess what she found next? [Hold up mouse. Let children answer.] Yep, a dead, half-eaten mouse and because she was in such a hurry and couldn't see it very well, she threw that in her basket. Time was running out. It was almost completely dark and Topsy needed to get home safe and sound before the owls came out. Why do you think she was afraid of owls? [Let children answer.] So she gave one last dig, and guess what she found? [Hold up acorns. Let children answer.] Yep, instead of some juicy, fat carrots all she found were some hard, dry acorns. And because she was in such a panic, she threw the acorns into the basket without thinking and ran all the way home. When she got home, Daddy Bunny was waiting. What do you think he said to her? [Let children answer.] And what did he find when he looked in the basket? [Let children answer.] What kind of stew would you get if you put a big, stinky bone, a half-eaten mouse, and a bunch of acorns in it? Would you like to have some for dinner tonight?

VOCABULARY:
Ambled
Decision
Delectable
Energy
Excuses
Halfheartedly
Loped
Parsnips
Pouch
Realized
Tasty morsel
Turnips
Speck

Making Shadow Puppets
Art

WHY we are doing this activity: to facilitate creative expression; to develop fine motor skills; to facilitate all components of language arts: reading, writing, speaking, and listening.

WHAT we will need:
Construction paper
Markers
Crayons
Pens
Popsicle sticks
Glue
Glue brushes
Preparation:
Scissors
Flashlight or other light source

HOW we will do it: To prepare, cut out a puppet shape from construction paper and glue a Popsicle stick on the back so that the paper doesn't flop over, and so that you have a handle for the puppet. On the back, write a few sentences. For example: "This is a little ghost who lives up in the attic." Lay the materials out on your project table. After you tell the children the previous shadow puppet story, show them the materials on the activity tables. Before they approach the project table, hold your puppet up, show the children where you placed the Popsicle stick and read the words on the back. Invite the children to make their own shadow puppets and to use the flashlight or light source to see what shadows their shapes make. The markers, crayons, and pens are for the children to color the cardboard shapes if they would like to. After the children have worked on them for a while, encourage them to give story dictation or to write their own words on the back of the puppets. Ask the children to read their words back to you, or if they prefer, read the words to them. Later in the day or week, gather together and have show-and-tell of the shadow puppets the children made.

Shadow Puppet Theater
Drama/Language

WHY we are doing this activity: to facilitate pretend play; to help children develop motor and speech coordination; to help children express themselves.

WHAT we will need:
Same theater arrangement as for
puppet story
Shadow puppets (made by children for
previous activity)
Shadow puppets from "Topsy and the
Carrot Patch" story
Experimentation box
Two flashlights
Duct tape

HOW we will do it: To prepare, use duct tape to tape two flashlights into the corners of the top of the experimentation box. Slant them at an angle so that the children can make shadows with their puppets at their own arm level. Encourage the children to use their shadow puppets in the theater and to make up stories with them, and also to use them in the experimentation box. What kind of shadows does each make? Are they different?

Shadow Match
Math (for three- and young four-year-olds)

WHY we are doing this activity: to facilitate matching exercise; to develop cognition through shape recognition; to help develop self-esteem and a sense of autonomy through use of a one-person work station.

WHAT we will need:
Distinctly different objects from the
classroom: manipulatives, toys, containers, etc. (many different shapes and
heights)

Black marker
Large piece of poster board
Container (to hold various objects)
"One person may be here" sign
(provided on page 165; copy and
enlarge for your won use)
Clear contact paper

Optional:
Strong light source

HOW we will do it: Spread your objects out over the poster board, lay them on their sides, and trace around each with a black marker. Color each outline in to make a "shadow" for every object, and then cover the poster board sheet with contact paper. Set the paper on the floor or a table, and put the objects in the container beside it. If feasible, use a strong light source to trace the actual shadow of each object.

During an Attention Getter time, show the children the game, and explain that if they like, they can match each thing to its shadow. Discuss the "One person may be here" sign and what it means.

Shadow Lengths and the Sun (Part 1)
Science

WHY we are doing this project: to help children understand that the earth moves around the sun and to help them understand how this affects our shadows.

WHAT we will need:
Large ball
Small piece of cardboard
Masking or packaging tape
Strong light or flashlight
Hot, sunny day
Butcher paper
Markers

Tip: Choose a day that is not windy for this activity, otherwise it will be difficult to keep paper in place while the children trace their shadows.

HOW we will do it: To prepare, cut out a piece of cardboard that is about 15 cm x 2.5 cm (6" x 1") and tape it to the ball so that the tab stands straight up and will cast a good shadow. Turn on your strong light or tape a flashlight to a high place so that it is shining down.

With the children, take the butcher paper and markers outside and notice the length of your shadows. Encourage your students to trace each other's shadows. Go outside again at noon and do this again. Compare the lengths of the traced shadows. Are they different? Invite the children to hypothesize about why this is.

Go back indoors, and show your students the ball and ask them to imagine for a minute that it is the planet Earth, and that the light is the sun. Ask them to notice the shadow of the cardboard. Move the ball around the light in an approximation of the way the earth moves around the sun. At a certain point the light should be exactly above the cardboard tab. Ask the children what they noticed about how the shadow changed. Let the children take turns moving the ball around the light if they want to. If possible, go out again in the late afternoon to trace and compare your shadows one more time.

Shadow Measure
(Part 2)
Science/Math

WHY we are doing this activity: to provide experience with standard measurements; to practice rational counting; to conduct scientific comparison of shadow lengths; to provide experience with writing down measurements in inches and feet; to help children cooperate with each other; to enable children to see visible proof that the earth moves around the sun, to encourage children to explore another facet of themselves and their bodies: their shadows. (For older preschoolers and kindergartners only.)

WHAT we will need:
  Three large pieces of butcher paper
  Markers
  Rulers and tape measures
  Sunny day

HOW we will do it: To prepare, gather a collection of rulers and tape measures that are child- user-friendly. For instance, are the centimeters marked in different colors than the meters? Are the numbers big? I have not succeeded in finding rulers or tape measures made specifically for children, but you can make your own by photocopying a ruler and using colored markers to write in bigger numbers. On top of one butcher paper piece print: "Shadow measurements at ___ in the morning." Write in the time you actually go outside. Make one for noon and one for late afternoon. Pin these charts up on the wall outside.

Before you go outside, as described in the previous activity, show the children the tape measures and rulers and explain what the measurements mean. When you go outside to look at your shadows, ask the children to find a partner who will measure their shadows for them. Help them read the rulers and tape measures and invite them or help them to write down the measurement on the chart. Do the same at noon and late in the afternoon. Compare the measurements. Remind the children of your demonstration with the ball, light, and cardboard tab, or do it again. Talk about the difference in shadow measurements and the reason for them.

What Is an Eclipse?
Science

WHY we are doing this project: to provide a hands-on activity in which children produce an eclipse; to develop self-esteem and a sense of autonomy through use of a one-person work station.

WHAT we will need:
  Two blocks (one large and one small)
  Strong flashlight
  "One person may be here" sign (provided on page 165; photocopy and enlarge for your use)

HOW we will do it: To prepare, arrange your light so that the object on the table will cast a shadow. Put the two blocks on the table and place the sign nearby. During an Attention Getter time, put the small block in front of the light, and ask the children if they see the block's shadow. Ask: "If I put the big block in front of the small one, do you think the small block will still have a shadow? If you would like to find out for sure, you can do this experiment yourself." Interpret or read the "One person may be here" sign together and talk about what it means. As the children explore the materials, talk to them about their discovery and use the word *eclipse*: "The big block *eclipsed* the shadow of the small block."

The shadows need to be cast on the table surface, so they can be traced. Set out the paper, objects, and markers on the table.

Ahead of time, take an object yourself and trace on paper the shadows you see when you put the same object in different positions. If you like, make another one and this time color the outlines in. During an Attention Getter time, show the children your different shadow outlines of the same object, and explain how you made it. Invite the children to make their own. Before they disperse to make their shadow outlines, make sure you say, "These are just how my outlines turned out. Everyone's will look different because we are all different individuals with our own ways of doing things." Pin up the children's pictures on your walls afterward.

Shadow Outlines
Science/Art

WHY we are doing this activity: to help children observe that when the position of an object is changed, its shadow is changed also; to promote creative expression.

WHAT we will need:
  Paper
  Markers
  Several flashlights or lights
  Objects (suggestions below)

HOW we will do it: Try to collect objects that will cast interesting shadows: fly swatters, latticed chairs, lattice-side plastic crates. Manzanita wood bird perches cast very dramatic shadows. If you know someone who has one, perhaps you can borrow it, or approach a pet store about doing so. Driftwood or gnarled branches can also achieve the same effect if they're positioned upright and secured in a base of wet sand or rocks. Arrange your lights and a long table so that several children can sit at the table and face the shadows cast by their objects. If you need to, use duct tape to attach flashlights to a surface like the edge of a shelf.

Shadow Walk
Gross Motor/Nature

WHY we are doing this activity: to sharpen powers of observation and to exercise gross motor muscle group.

WHAT we will need:
  Sunny day
Optional:
  Rope

HOW we will do it: Tell the children that we are going to go on a walk to see how many interesting shadows we can find. Use the rope for everyone to hold on to. As you take your walk, see how many interesting or pretty shadows you find, for example: the shadows cast by a chain-link fence, lattice railing, or iron gate, very small shadows (blade of grass), or very large shadows (tall building). What happens when a cloud moves in front of the sun? When you get back to home or school, talk about the shadows you saw.

Musical Shadow Tag

Music/Gross Motor

WHY we are doing this project: to exercise the large muscle group; to develop appreciation for music; to develop listening skills.

WHAT we will need:
   Tape recorder
   Music
   Hot, sunny day
   Grassy area

HOW we will do it: Play a game of musical shadow tag. The children run and dance to the music until it stops—then they must jump onto a shadow and freeze. It can be the shadow of a stationary object, or the shadow of another person. After the children learn the game, let them take turns turning the tape recorder on and off; however, young children tend not to allow much time in between so encourage them to wait each time, so that the other children have a chance to run and dance.

Literature

Symbol Key: *Multicultural
              +Minimal diversity
              No symbol: no diversity or no people

Anno, M. (1976). *In shadowland*. New York: Orchard Books. (This is an excellent story book.)

Bulla, C. R. (1994). *What makes a shadow?* New York: Scholastic.

Cendrars, B. (1982). *Shadow*. New York: Charles Scribner's Sons. (This is a translation of an excellent story taken from conversations with African shamans.)

Goor, R. & Goor, N. (1981). *Shadows here, there and everywhere*. New York: Thomas Y. Crowell.+

Gore, S. (1989). *My shadow*. New York: Doubleday.+

Simon, S. (1985). *Shadow magic*. New York: Lothrop, Lee & Shepard Books.*

Extenders

Science: After you conduct the Shadow Lengths and the Sun experiment, put out a flashlight and a tall object on a tray on the floor. Encourage the children to raise the flashlight over the object from one side to the other. What happens to the shadow?

Science: Set up two lights to shine on an object. How many shadows are there? If you can, set up more than two lights to shine on an object and invite the children to count the shadows again.

Math: Count the shadows in your room. Do it as a group, or invite the children to do it by themselves. If you have blinds on your windows, compare how many shadows you see when the blinds are open and when they are closed.

Language: After you count the shadows in the room, make a language chart of all the shadows. Write down the name of each object that made a shadow and invite the children to draw a picture of them. Afterward, read the chart together.

MARVELOUS MOTION

Attention Getter: Put some ball bearings or marbles in tins with lids. When the children are gathered, let them take turns moving the tins. Are the objects inside still or moving? How can they tell? If you like, have the children guess what is inside the tins and then let them take the lids off. Explain that when something is moving, it is in *motion*. Say the word *motion* several times as you pass the tins and ball bearings around the group so that each child can feel the ball bearings roll around in the containers. Ask the children to guess what you will be talking about and working with during the next few weeks.

Safety Precaution: Supervise children closely during all activities involving marbles.

Marble Mania

Science

Square or rectangular boxes
Activity sign (provided below; photocopy
and enlarge for your use)

WHY we are doing this experiment: to help children experiment with centripetal and centrifugal force; to develop reading skills through an activity sign.

The facts of the matter: *Centripetal force* is the force that pushes an object toward the center of a circle. *Centrifugal force* is the force by which an object pushes outward, away from a center. With marbles and bowls, the centrifugal force causes the marbles to climb to the widest part of the bowl, the sides at the top, which is also the part farthest from the center of the circle the marbles are moved in when the bowl is manually moved around and around.

WHAT we will need:
    Ball bearings or marbles
    Large plastic bowls

HOW we will do it: To prepare, post the activity sign on the wall near the spot where your materials will be. Put all materials out on the activity table. Marbles are going to be rolling everywhere during this activity, but children do enjoy it. If you want to try to contain the marbles, designate a particular area for this experiment and, if you like, try to block it off a little with furniture, or by taping cardboard box borders around the perimeters.

During an Attention Getter time, read/ interpret the activity sign together. Ask the children to predict how the marbles will react when the bowls are rolled around. Ask them if they see anything in the room that will help them conduct this experiment. Invite them to do so. Why do the marbles climb the sides of the bowls and shoot out? (The sides of the

Marble Experiment:

Put some marbles in a bowl, and move the bowl around in a circle. What happens to the marbles?

bowls push the marbles toward its center, but the weight of the marbles push them out toward the sides of the bowls. When the children move the bowls around and around, this makes the marbles move away from the bowls' centers, too.) Does the same thing happen when the marbles are rolled in the square or rectangular boxes?

More Centrifugal Force
Science

WHY we are doing this experiment: to provide children with more hands-on experience with centrifugal force.

WHAT we will need:
    Small buckets with handles
    Water
    Paper towels
    Hot, sunny day

HOW we will do it: Encourage the children to fill the buckets half full of water, and to swing them over their heads, vertically, as hard and as fast as they can. Does the water fall out? Remind the children that centrifugal force pushes things out, away from the center of the circle. The force of the circular movement they make with their arms forces the water out, toward the bottom of the buckets. Have paper towels on hand in case there are some accidents!

How Do Ball Bearings Work?
Science

WHY we are doing this project: to enable children to discover, through hands-on experimentation, that friction caused by rolling is less than sliding friction; to develop self-esteem and a sense of autonomy through use of a one-person

work station; to develop a sense of mechanical competence; to develop reading skills.

WHAT we will need:
    Four coffee cans (all the same size)
    Ball bearings (all the same size,
        about twelve)
    Red and brown contact paper (or con-
        struction paper)
    Activity sign (provided on page 54;
        photocopy and enlarge for your use)
    "One person may be here" sign
        (provided on page 165; photocopy and
        enlarge for your own use)

HOW we will do it: To prepare, make both signs and pin them up on the wall near the work table. For this experiment, you can use ball bearings (available in hardware stores) or steel marbles (available in toy stores). Ball bearings in hardware stores vary in size and price; typically, they range from $.07 to $.28 each.

Cover two of the cans in brown contact or construction paper and the other two in red. Turn one brown coffee can upside down on the table, and line the ball bearings around the rim. Put the other brown coffee can on top of the ball bearings, right side up. Turn one red coffee can upside down on the table and put the other red coffee can on top of it, right side up.

During an Attention Getter time, read/interpret the signs together. Ask the children to predict which can will be easiest to slide and move over the can underneath. As the children take turns experimenting, ask them about the results. Why is the coffee can on the ball bearings so much easier to turn and slide? Rub your hands together and ask the children if they remember that two things rubbing together make *friction*. There is much less friction when one coffee can rolls on ball bearings than when two coffee cans rub against each other, because the ball bearings roll. Ball bearings are often used in cars and other machines that move.

Slide one red coffee can over the other.

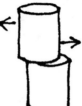

Slide one brown coffee can over the other.

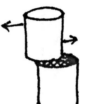

Which one slides easiest? Why?

Kinetic Energy
Science

WHY we are doing this experiment: to enable children to experiment with kinetic energy through hands-on exploration.

Definition: *kinetic*: of or resulting from motion

WHAT we will need:
Cardboard cylinders (from cling
  wrap boxes)
Marbles (large and small)
Double-sided tape
Blocks

HOW we will do it: To prepare, cut the cardboard cylinders in half, lengthwise. Use blocks to make a number of ramps with a wide variety of lengths and heights. Each ramp should consist of a long flat part, secured to the floor with double-sided tape, and a lifted ramp, secured on one end to the flat ramp, and on the other end to the blocks that support it. Use double-sided tape for this purpose also. Line up four or five marbles in the end of each ramp. Reserve one marble near the top of each ramp.

During an Attention Getter time, put one marble at the top of a ramp. Ask the children: "If I let this marble roll down the ramp, do you think anything will happen to these other marbles? What do you think it will be?" Encourage the children to express their predictions. Put the marble back down on the floor without dropping it down the ramp, and invite your students to use the materials to find out if their predictions are true. As they roll marbles down the ramps, what happens to the other marbles sitting at the bottom? Why?

The facts of the matter: Because the marble that rolls down the ramp is moving, it has energy. When the rolling marble hits the marbles that are standing still at the end of the ramp, the energy of the rolling marble is passed from marble to marble until the one at the very end is pushed off the ramp and rolls away. The still marbles at the end of the ramp move because of the energy, or motion, of the rolling marble. This is an example of kinetic energy.

Developmental differences: Three- and young four-year-olds will enjoy rolling marbles down the ramp. Conduct the kinetic energy experiment yourself and see if any children become interested. Older children will also enjoy using the ramps and marbles, as well as conducting and observing the results of the experiment.

Marble Count
Math

WHY we are doing this project: to use motion to practice rational counting and subtraction; to develop self-esteem and a sense of autonomy through use of a one-person work station.

WHAT we will need:
> Marbles or steel ball bearings
> (large and small)
> Two coffee cans (one large with plastic lid and one small)
> Long cardboard cylinders (from cling wrap boxes)
> Brown and red contact paper (or construction paper)
> Twine, books, or blocks (to hold up ramps)
> Writing sheets (provided on page 56; photocopy and enlarge for your use)
> Pens
> Blank paper
> "One person may be here" sign (provided on page 165; photocopy and enlarge for your use)

Preparation:
> Scissors
> Double-sided tape

Glue
String
Putty
Exacto knife

HOW we will do it: To prepare, cut the cylinders in half, lengthwise. In this activity, the children are going to count marbles as they roll them down the cylinder ramps into a coffee can. (The falling marbles make a very satisfying clunking noise as they hit the metal of the can.) Put all the marbles in the small can. Use the Exacto knife to cut a hole in the plastic lid of the large can. Make the hole big enough for marbles to pass through. Use the blocks or books to prop up the ramp at an angle, and use putty or double-sided tape to secure the lower end over the coffee can hole.

You can make the ramp as long as you wish. It is fun to make a super-long one, but to do this you need adequate space to create an effective angle. Glue the ends of the cardboard ramps together. You can either hang your ramp with string secured to the ceiling with tacks, or prop the ramp up with blocks. You may need to use putty or double-sided tape to help secure the ramps.

During an Attention Getter time, show the children all the marbles, and ask them to predict how many there are. Look at the writing sheets and together, read/interpret them. Show the children the pens, and let them know they can also use the blank paper to record the marbles they count, if they would like. Talk about the "One person may be here" sign.

Developmental differences: Three- and young four-year-olds will be most interested in rolling the marbles down the ramps, into the cans. If you like, count their marbles out loud while they do this. They may use the blank or writing sheets for scribbling. Older children are likely to count the number of marbles rolled down the ramp and to record the number they count.

Roll the marbles into the brown can. How many marbles do you count?

Take three marbles away. How many marbles are left?

Powerful Pulleys
Science/Fine Motor

WHY we are doing this project: to enable children to discover that the energy of one motion can be harnessed to cause another motion; to develop fine motor skills; to provide experience with manipulatives.

WHAT we will need:
  Wood (flat, square piece about
    25 cm x 25 cm [10" by 10"])
  Spools
  Finishing nails (available at
    hardware stores)
  Strong rubber bands
Preparation:
  Hammer
  Pliers
  Pencil

HOW we will do it: For this project, you will need to make a pulley board. The spools will rotate on finishing nails, which have been hammered into wood and are connected to each other by rubber bands. When the children turn one spool, the others will turn also.

Stretch the rubber bands and decide how far apart the spools should be. Mark the places with pencil. The rubber bands, when stretched around two spools, should not be so tight that the rubber bands are in danger of breaking, but they must be snug enough to function. Hammer a finishing nail into each spot you marked with the pencil. Finishing nails are better for this project than regular nails, because the heads are minimal; this enables children to slide the spools on and off the post. This way, they can rearrange the pulley board in any way they would like. This provides a good, fine motor exercise. On two, three, or more spools, hammer one nail into the top of the spool, and use the pliers to bend it so that it can be used as a handle to turn the spool. Make as many of the pulley boards as you wish, and put them out for the children to explore. (They love them!) As the children use the boards, ask them about what they see happening.

More Pulley Play
Science

WHY we are doing this project: to enable children to discover that a pulley makes lifting easier; to develop reading skills with an activity sign.

WHAT we will need:
Wire clothes hanger
Spool
Strong twine (or thin rope)

Two small buckets with handles
Potatoes or rocks (six, roughly the
    same size)
Plastic margarine tub lid
Activity sign (provided below; photocopy
    and enlarge for your use)
Red and black construction paper
Tape
Preparation:
Scissors or Exacto knife
Pliers

HOW we will do it: To prepare, unbend the hanger and thread the spool onto it. You may need the pliers to help you untwist it, and then twist it back again.

Find a high place to hang the hanger. Put three potatoes in each bucket. (You do not want the weight of the bucket to pull the hanger out of shape, so only put in as many potatoes as the hanger will accommodate without bending.) Take a length of strong twine (or thin rope), and

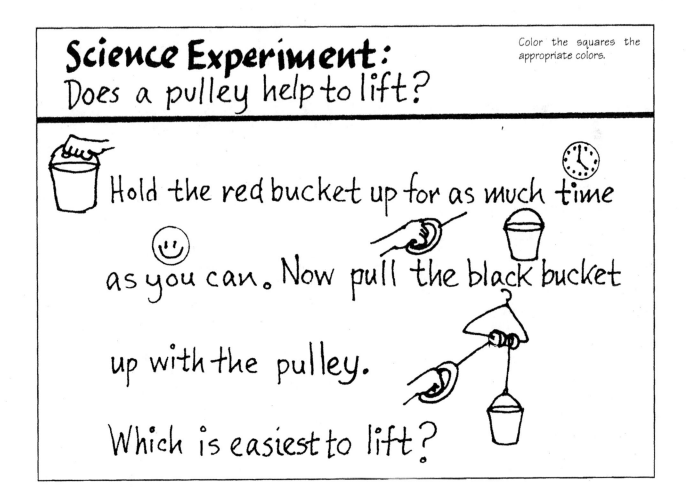

Science Experiment:
Does a pulley help to lift?

Color the squares the appropriate colors.

Hold the red bucket up for as much time as you can. Now pull the black bucket up with the pulley.

Which is easiest to lift?

tie one end to one bucket while it is standing on the floor. The twine should be long enough so that the other end is easily reached by children when the twine is threaded over the spool.

Cut the center out of the margarine lid. This is going to be the pulley handle, attached to the other end of the twine. If you can find something (a plastic ring from a game or toy) that is smoother to hold, use that instead. Line up the twine or rope so that it lies on the spool or hanger. (See activity sign)

Cut out two large squares from construction paper (one red and one black). Tape the black one to the bucket attached to the pulley and the red square to the bucket that is standing alone. Put the red bucket of potatoes beside the pulley. Post the activity sign on the wall near the pulley.

During an Attention Getter time, ask the children if they know what a pulley is. Show them the one you made, and together, read/interpret the activity sign. Ask your students to predict which bucket of potatoes will be easiest to lift. When the children explore the materials, ask: "Which bucket of potatoes is easiest to hold up for a long time—the one with the pulley or the one without? Why do you think that is so?" Discuss. (The rolling pulley does part of the work of lifting. Pulleys are used in building. Pulleys help lift large bundles of bricks and heavy planks. Pulleys are also used to unload cargo from ships.)

Levers Also Help Lift
Science

WHY we are doing this experiment: to help children understand that a lever lifts; to allow them to discover, through hands-on experimentation, that when they push down on a lever, an object is lifted up; to help children understand that the point of support on which a lever turns is called the *fulcrum*.

WHAT we will need:
Rulers (small ones, 15 cm and 30 cm [6" and 1'])
Meter or yardsticks

Blocks
Planks from block sets
Books
Twine
Pencils
Duct tape
Cylindrical objects of diverse diameters (e.g., sturdy cardboard tubes, plastic containers, cylindrical blocks)
Demonstration:
Two colors of construction paper
Little people pattern (from shipwreck activity in Soap Science unit)

HOW we will do it: To prepare, use the twine to tie bundles of books and blocks together. The bundles should vary in size and weight, so use small books as well as large, and paperbacks as well as hardcovers. Do the same with blocks. One bundle can consist of one block by itself.

Make a variety of levers by balancing small rulers on pencils or other thin, long, cylindrical objects. Be sure the ruler moves smoothly on the fulcrum or the cylindrical object underneath it. Put yardsticks and planks on the other larger cylinders. In addition, make some levers which use blocks instead of cylinders for fulcrums.

Match the bundles to the levers according to the strength of the plank or ruler and the weight of the bundle. Tie the bundles onto the levers so they do not fall off when the children lift them. If necessary, use duct tape to help secure them. Nearby, set out loose rulers, yardsticks, planks, cylinders, blocks, strips of duct tape, and lengths of twine, so that the children can set up their own levers and bundles.

Use the little people patterns (from the shipwreck activity in the Soap Science unit) to cut out two paper people, each one from a different color of paper. For a demonstration at an Attention Getter time, create a small seesaw with a cylinder and a ruler, and tape a paper person on each end.

During an Attention Getter time, ask the children if they have ever been on a seesaw. Push down on one end of your miniature seesaw and say: "When I push down on this end, what happens?" (The other end rises.) Say: "This is a lever. I push down on one end of the lever, and it lifts the other end." Point to the cylinder on which the lever rests and at the point on which the lever balances, and say: "This part is called the *fulcrum*. The fulcrum is where the lever turns." Say the word several times together as the children take turns pushing down on the miniature seesaw.

Show the children the other levers you set up, and ask them to predict whether or not they will be able to lift the bundles with them. Invite the children to explore the materials. Is there a difference between how a lever on a cylinder works with how a lever on a block works? Which levers lift faster and more easily: those underneath light bundles or heavy bundles? Invite the children to use the loose materials to make their own levers.

Momentum Magic
Science/Gross Motor/Sensory

WHY we are doing this experiment: to enable children to experiment with momentum and to develop the large muscle group.

Definition: *momentum:* the impetus of a moving object, or, a strength or force that keeps growing.

WHAT we will need:
Large plastic bags with handles
Rocks (large, medium, and small-sized)

HOW we will do it: Put several heavy rocks in each bag. Encourage the children to put the bags down on the ground and to then twist the plastic handles many times. Ask them to predict what will happen when they pick the bags up again. Invite them to do so. What happens? Are the bags easy to stop once they start untwisting? Do the bags unwind quickly or slowly? What happens when the bags are filled with medium- or small-sized rocks? Experiment with these variations. (The weight of the rocks creates *momentum* as the bags unwind. The weight of the rocks creates a force or motion that grows and grows and is hard to stop.) You may want to repeat this activity at the next day's Attention Getter time, and have the children repeat the word *momentum* several times as their bags unwind.

More Momentum
Science

WHY we are doing this experiment: to enable children to discover that the motion of the fluid inside a raw egg creates momentum.

WHAT we will need:
Raw eggs
Hard-boiled eggs
Small bowls
Preparation:
Marker

HOW we will do it: Mark each hard-boiled egg with a marker so that the cooked eggs are easy to identify. Put all materials out on the activity table.

During an Attention Getter time, ask the children if they remember what *momentum* means. (Movement that keeps growing.) Invite the children to hold the hard-boiled eggs and to guess whether they are raw or cooked. Do the same with the raw eggs. Ask your students to predict whether there will be a difference in the way that a raw egg spins in a bowl, and the way that a cooked egg spins. Before the children conduct this experiment, talk about how the raw eggs should be handled to keep them intact.

As the children explore the materials, ask them what they notice about which eggs are easiest to stop, once they are in motion. Why do they think this is so? (The fluid inside the raw eggs creates momentum. When the children try to stop the raw eggs from spinning, the fluids inside continue spinning and keep the egg in motion.) Which egg is easiest to spin from a standstill? (The motion of the shell has to be transferred to the fluid of a raw egg, which is why a cooked egg is easier to stop spinning.)

Developmental differences: Three- and young four-year-olds: The temptation to break the eggs will be too much for this age group. Be prepared to facilitate a sensory exploration of both kinds of eggs. Conduct the momentum experiment yourself and see if any children become interested. Older children may be inclined to break open the eggs, but are also likely to be interested in conducting the momentum experiment with eggs that are still intact.

Squeeze Bottle
Science

WHY we are doing this experiment: to enable children to discover that one motion can create a series of motions.

WHAT we will need:
> Plastic squeeze bottles
> Tissue paper
> Small paper cups
> Small containers of water
> Medicine droppers

HOW we will do it: Before you facilitate this unit, start gathering the plastic bottles. They should be the kind that are made out of soft, flexible plastic so that the children can easily squeeze them hard. Put all materials on the activity table.

When the children are gathered during an Attention Getter time, pass the squeeze bottles and encourage the children to squeeze them into their faces. Can they feel the air? What makes it shoot out of the bottles into their faces? (The squeezing motion of the children's hands.) Show your students the materials on the activity table. Sprinkle drops of water from a medicine dropper onto a tissue, and then roll it up into a plug. Pack it into the top of the bottle and put one of the small paper cups on top, upside down. Ask the children to predict what will happen if the bottle is squeezed hard. Instead of demonstrating this, encourage the children to prepare their own experiments and find out.

As they conduct the experiment, discuss the results and the sequence of motions. (Squeezing the bottle makes the air shoot up, which forces the wet tissue paper out, which makes the cup fly off. One motion causes another motion which causes another motion.)

Air and Water in Motion
Science/Sensory

WHY we are doing this experiment: to enable children to discover that with enough force, air in motion can displace water; to facilitate cooperation between children; to develop hand-eye coordination; to provide a sensory experience.

WHAT we will need:
> Sensory table
> Tubs
> Water
> Plastic soda bottles (16 oz.)
> Bendable straws
> Coffee scoopers
> Funnels

Optional:
> Food coloring
> Glitter (or colored sand)

HOW we will do it: To prepare, set up the tubs of water, and put the bottles inside. If you'd like, add food coloring and glitter to the water.

During an Attention Getter time, fill a bottle with water, put a bendable straw inside the bottle, and then turn it upside down and hold it partway under the water.

Ask a child to hold the bottle upside down in position while you hold the straw. Ask the children: "What do you think will happen if I blow through the straw as hard as I can?" Encourage the children's predictions. Say: "I had to ask Miguel to hold the bottle for me so that I could think about holding my straw and blowing into the bottle. What could you say to someone to ask them to hold the bottle when you want to try this science experiment?" Encourage the children to express different ways of asking for help. Also talk about the fact that everyone should use their own straw so that germs are not spread.

As the children use the straws to blow into the upside-down bottles in water, encourage their observations. Why does the water not stay in the bottle? What is pushing it out? What does this experiment tell us about whether air can move water?

Pendulum Play

Science

WHY we are doing this experiment: to introduce children to the motion and principles of pendulums.

WHAT we will need:
    String
    Duct tape
    Washers (various sizes and weights)
    Jumbo paper clips
    Objects of different weights and lengths
        that can be hooked onto the paper
        clips: (e.g., pieces of chain bracelets,
        chain links, key rings with and without
        keys, unattached keys)

HOW we will do it: To prepare, unbend the paper clips so that they can be used as double hooks. Cut a variety of lengths of string and tie one length to one hook of each paper clip, leaving the other hook free to hold washers. Cut strips of duct tape and stick them onto the edge of the activity table. Put out the washers and the prepared paper clips as well.

Make a sample pendulum by taping one end of the string to the edge of the table and putting an object on the hook on the other end as it hangs down. During an Attention Getter time, ask one of the children to set the pendulum in motion. As the pendulum swings, say: "Back and forth and back and forth," in time to the pendulum's swing. Encourage the children to chant this with you and notice how the distance covered by the pendulum becomes less and less as it slows down. Tell the children that the swinging time of a pendulum is called a *period*.

Show your students the materials on the activity table and encourage them to make their own pendulums. Ask the children about what they discover. Does a pendulum of heavy objects take the same time to swing as one with lighter objects? Does a pendulum with a long string take the same time to swing as one with a short string? What difference does it make when the object at the end of the pendulum is long, like a key or bracelet chain, instead of compact like a washer? Is the motion of the key or bracelet the same as the pendulum's?

Pendulum Rhyme

Music/Movement/Rhythm/Gross Motor

WHY we are doing this activity: to build on the experience in the previous activity; to help children enjoy their singing voices; to reinforce the motion of a pendulum through chanting; to help children feel rhythm.

WHAT we will need:
    Lengths of string (about 37.5 cm [15"])
    Washers
    Rhyme: "PENDULUM, PENDULUM"

"PENDULUM, PENDULUM"
Pendulum, pendulum
swinging to and fro,
first it swings so very fast
and then the swinging slows.

HOW we will do it: Make a pendulum for each child by tying a washer to each length of string. During an Attention Getter time, hand each child a pendulum and start them swinging. Say the rhyme in time to the swinging. Experiment with saying the rhyme quickly and slowly. To make the pendulum swing very fast while you chant quickly, hold the pendulum close to the weight at the end. To make the pendulum swing slowly, hold the pendulum at the top of the string. Stop moving your hand on the last word of the chant and notice how the arc of the pendulum diminishes and eventually slows down to a standstill. When you say the chant quickly and slowly, can everyone keep time together and stop at the same time? As a variation of the chant, sing the words in the tune of the first few lines of the "ABC" song.

For a gross motor exercise, stand up, hold your hands together, and swing your arms as pendulums, in time to the song or rhyme. Try it with both hands loose. Try hanging your head down low and swinging your whole torso like a pendulum. Stand on one leg and swing the other one like a pendulum as you sing the song. Be creative!

Pendulum Pair

Science

WHY we are doing this experiment: to enable children to observe how one pendulum transfers its motion to another.

WHAT we will need:
- String lengths (all the same length)
- Washers
- Duct tape
- Sticks (at least 17.5 cm (7") long)
- Jumbo paper clips

Later part of experiment:
- String in a variety of lengths
- Objects of different weights to hang on pendulums

HOW we will do it: Unbend the paper clips and tie these hooks to one end of each string length. Arrange all materials on the activity table as specified in the Pendulum Play activity. The weight at the end of the pendulum is called the *bob*. Make a pendulum pair by taping two pendulums with equally weighted bobs to the edge of the table, about 15 cm (6") apart.

During an Attention Getter time, take a stick and wrap each pendulum string around the stick once about halfway up each string. Ask the children what will happen if you just swing one bob. Will the other pendulum swing too? Rather than demonstrate this, let the children make their own pendulum pairs to find out. What happens? Because they are connected, the motion of the first bob affects the second one and sets it swinging, too. Encourage the children to take the experiment in other directions, by providing a variety of string lengths and differently weighted bobs. What happens if the stick is moved down closer to the bobs? What happens if the bobs are different weights? What happens if the string lengths are different, but the pendulums are still connected? Find out!

Air Boats

Science

WHY we are doing this project: to enable children to discover Newton's third law of motion (for every action, there is an equal and opposite reaction); to provide a novel way for children to experiment further with the force of air and the motion it causes.

WHAT we will need:
- One-quart milk cartons
- Good quality balloons
- Sensory table filled with water (or children's pool)
- Chip clips (clips that keep bags of snack foods closed)

Preparation:
- Exacto knife or razor blade

HOW we will do it: To prepare, cut the milk cartons in half, lengthwise. Glue the pouring spouts back together, and lay each carton down on a table, open side up, to simulate a boat. The top triangular parts of the cartons form the prow of the boats. In the opposite rim of each carton, use an Exacto knife or razor blade to cut a small hole in the middle. It should be just big enough for the opening of a deflated balloon to be pushed through.

Put all materials out near the sensory table of water. During an Attention Getter time, blow up one of the balloons that has been inserted in a boat, and put a chip clip on the end to keep it from deflating. Set the boat in the pool or sensory table of water, and ask the children what they think will happen when you take the clip off. Rather than demonstrate this, encourage the children to prepare their own experiments. Help young preschoolers push the balloon openings through the hole in the milk cartons and blow up the balloons. Either help them pinch the ends of the balloons with their fingers until they are ready to release them, or use chip clips for this purpose. What happens to the boats?

Discuss with the children the "push/push back" law of motion. When there is a push in one direction, it makes a push in the other direction. The air blasts out of the balloon in one direction and makes the boat shoot off in the opposite direction. To test this theory, encourage the children to put their fingers near the end of the balloon after the air is released. In what direction is the air moving? Is it the same direction as the boat?

each have their own and germs are not spread. As the released air from the balloons moves the rockets along the yarn, ask the children if they remember the push/push back law. Encourage them to put their fingers near the opening of the balloon when the air is released. In what direction is the air moving? Does the straw move in the same direction? Discuss.

More Push/Push Back
Science

WHY we are doing this experiment: to give children another opportunity to explore the push/push back law.

WHAT we will need:
> Yarn
> Masking tape
> Strong balloons
> Straws
> Chip clips
> Marker

HOW we will do it: To prepare, string yarn across a long stretch of the room, at a height the children can easily reach. Fix one end securely to the wall. Wind masking tape tightly around the other end of the yarn so that the children can easily thread a straw onto the yarn. Tear off masking tape strips and stick them on the edge of the activity table.

Prepare a rocket: Thread a straw onto the yarn line, inflate a balloon, and tape the balloon to the straw.

Use a chip clip to close the end of the balloon after it is inflated. Ask the children what they think will happen when you take the chip clip off. Rather than demonstrate this, encourage your students to prepare their own experiments. Help young preschoolers inflate their balloons, secure the openings with chip clips, tape the straws onto them, and thread the straws onto the yarn. Use the marker to write children's names on their balloons so that they

The Book of Motion
Language/Art

WHY we are doing this project: to develop all components of language arts: reading, writing, speaking, and listening; to facilitate artistic expression.

WHAT we will need:
> Construction paper
> Magazines (any catalogs, home-improvement magazines, or mechanics publications with photographs of moving parts and machines)
> Glue
> Small containers for glue
> Glue brushes
> Children's scissors
> Crayons
> Markers
> Pens

Preparation:
> Stapler

HOW we will do it: Begin collecting the magazines several weeks ahead of time. Ask friends, neighbors, family, and parents to help you collect them.

Tip: If you see any multicultural, anti-bias photographs in the magazines as you do this project, save them for the Repair Shop activity that follows.

To prepare blank books, lay several pieces of paper on top of each other and fold them over. Staple along the fold. Put all materials out on the activity table.

Make a sample motion book. Write a title on the cover, and "by (your name)." Cut out pictures that interest you of moving parts and machines and glue them into your book. Write a few words about each. Mine is as simple as this: "This is a grandfather clock. It has pendulums a little bit like the ones we played with in class." "This is called a swivel chair. If you sit on it, you can turn yourself around and around on it." "This machine is a bulldozer. The big shovel moves up and down. The person who operates a bulldozer can move the shovel by pushing or pulling a handle. Sometimes I see bulldozers at work when I'm driving." Draw a picture of a moving machine of your own invention. For example, the words under my invented machine say: "I don't know what this machine is, but it has a spring that would be fun to press down. Then the spring bounces back up." Be creative and make your book however you'd like.

During an Attention Getter time, show the sample book to the children. Ask them if they see anything in the room that would help them make their own sample books. Mention that everyone's book will look different because we all have our own ways of doing things. Take story dictation, support invented spelling, spell words, or write them down to be copied, depending on your children's needs.

When the children are finished making their books, have a few children every day show-and-tell their words and pictures.

Repair Shop
Dramatic Play/Multicultural/Anti-Bias

WHY we are doing this project: to promote child-to-child interaction; to provide children with the opportunity to work out real-life emotions through fantasy play; to develop mechanical competence; to create a multicultural, antibias play area; to expand vocabulary; to develop speaking and listening skills.

WHAT we will need:
 Broken machines with moving parts

(e.g., clocks, watches, cameras)
 Toy screwdriver and screw sets
 Gear and sprocket toys
 Toy cash register
 Play money
 Signs (provided on page 65; photocopy
  and enlarge for your use)
 Blank paper
 Receipt books
 Pens
 Telephones
 Butcher paper
 Markers

HOW we will do it: Color in the signs provided and pin them up on the walls of your repair shop. Look for multicultural, anti-bias photographs from magazines and put these up on the walls as well. Set up all the materials to simulate a repair shop. If you are using broken machines like clocks and cameras, make sure you remove any rusted or jagged parts.

Pin the butcher paper up on the wall in the repair shop, and during an Attention Getter time, ask the children what the name of the repair shop should be. Print the name on the butcher paper. Invite your students to color and decorate it. Together, read the signs you have posted around the walls. Ask the children what they could use the blank paper and markers for. Invite them to explore the materials. To encourage fantasy play, call the shop to ask how much a repair would cost, or bring something in to the shop to be fixed.

Paint in Motion
Science/Art

WHY we are doing this activity: to allow children to use motion to create art and to facilitate artistic expression.

WHAT we will need:
 Plastic mustard squeeze bottles
  (with adjustable nozzles)
 Paint
 Newspaper

Yarn (or twine)
Thick, strong, rubber bands
Tacks
Newspaper
Large pieces of paper
Duct tape
Exacto knife or scissors
Tub of soapy water
Paper towels
Vinyl tablecloth

HOW we will do it: In this project, you are going to hang bottles of paint upside down. The bottles will be supported by harnesses, made of rubber bands and string, and will swing around when moved. Begin collecting your mustard bottles several weeks ahead of time. In order to control the flow of paint, you need bottles with screw tops, which gradually enlarge the opening.

To prepare, use the Exacto knife or scissors to cut out the bottoms of each bottle so that you can easily refill them with paint when they are in the harnesses. Do not cut off much more than the very bottom. Deep sides prevent paint from splashing out when the bottle is swinging. Twist a rubber band very tightly around each bottle, and tie three or four lengths of string around the band to make a harness.

Secure the strings to the ceiling with tacks. Make sure the bottles are hanging low enough for the children to push them.

Spread the vinyl tablecloth on the floor underneath the bottles and, on top of that, many layers of newspaper. Make sure the openings are closed when you fill the bottles with paint, and make sure the paint is neither too thick nor too runny. When you are ready to facilitate the activity, put paper underneath the bottles, unscrew the openings slightly, and invite the children to swing and push them. When the pictures are dry, hang them up on the walls.

Literature

Symbol Key: *Multicultural
+Minimal diversity
No symbol: no diversity or no people

Ardley, N. (1992). *The science book of motion.* San Diego, CA: Harcourt Brace Jovanovich.*

Barton, B. (1987). *Machines at work.* New York: Thomas Y. Crowell.*

Burton, V. L. (1939). *Mike Mulligan and his steam shovel.* Boston, MA: Houghton Mifflin.

Burton, V. L. (1943). *Katy and the big snow.* Boston, MA: Houghton Mifflin.

Lafferty, P. (1992). *Force and motion.* Boston, MA: Dorling Kindersley Books. (This book has excellent photographs which you can look at and discuss with the children.)

Rockwell, A., & Rockwell, H. (1972). *Machines.* New York: Macmillan.

Extenders

Science/Fine Motor: Introduce the children to tiddly winks and spinning tops. What does it take to make a tiddly wink hop up into the air? Does a spinning top always spin in the same spot?

Gross Motor: Take a walk outside and see how many moving parts of machines you can spot (e.g., wheels of cars, cranes, drills, and jackhammers). How much natural motion (e.g., wind blowing grass, leaves, and puddles) do you see? If you live or teach near a mall, walk through it and see how many moving machines or parts you spot (e.g., exercise bicycles, swinging or revolving doors, swinging pendulums, turning clocks or music boxes).

Math: When you experiment with the air boats and rockets, measure the distance they travel. Inflate the balloons to different degrees, mark the place where each boat/rocket stops moving (after the air is released), and then measure. Is the amount of air in the balloon related to the distance the rockets and boats travel?

MAGNIFICENT MAGNIFIERS

The facts of the matter: A convex lens bends light as the light enters and again as it leaves. This bending of light magnifies objects underneath the lens.

Although young children may not grasp the concept of bent light, their hands-on exploration of magnifiers in the activities below will lay the foundation for later understanding.

Several weeks ahead of time, start collecting a variety of glass jars and bottles, as well as clear plastic containers.

Throughout the unit, talk to the children about the glass in magnifying glasses, and about how important it is to use magnifiers carefully because they are breakable. Plastic magnifiers can be purchased; however, the ones I have seen have not been of very good quality in regard to their magnifying ability. Choose the magnifiers that you feel comfortable using. Also, let the children know that when you use glass jars in an experiment, the glass jars must not be moved.

Attention Getter: Pass a magnifying glass to every child. Ask them to examine their fingertips underneath the magnifier. How do their fingertips look different? Suggest that the children examine the palms of their hands. Ask the children to find a partner. Have the partner hold a magnifying glass over one palm. Compare the magnified palm with the one that is not, and ask your students what they notice. Have the children switch roles so that everyone gets a chance to make this comparison. Then show the children how to hold a magnifying glass a few inches away from their friends' faces, and examine eyes, nose, and mouth. Explain that when light passes through the magnifying glass, it bends, and then it bends again when it leaves. This makes everything under the glass look bigger, or magnified. Ask the children to guess what you will be working with over the next few days.

Fingerprint Fun
Sensory/Science

WHY we are doing this project: to help children experiment with magnifying glasses; to provide a sensory experience; to enable children to discover more about their bodies; to facilitate a group project.

WHAT we will need:
  Magnifying glasses
  Ink pads (nontoxic pads)
  Butcher paper
  Masking tape

HOW we will do it: Spread the butcher paper out on your activity table and tape down the edges with making tape. Set out the ink pads and magnifying glasses. During an Attention Getter time, approach the table and make a thumbprint. Then examine it with a magnifying glass, and encourage the children to make their own explorations. Together, discuss what you notice. Use the words *magnified* and *magnifying glass* during your discussion.

Science kits: Collect some boxes or tins with lids, and several nontoxic ink pads. Put an ink pad, sheets of paper, and a magnifying glass in each box or tin. Put these science kits out for the children to explore independently. Discuss with the children that ink should be pressed only onto the paper and not on furniture or walls.

Water Magnifying Activity (Part 1)
Art/Sensory

WHY we are doing this project: to facilitate creative expression and to provide a sensory experience.

WHAT we will need:
  Tempera paints (vivid colors)
  Pieces of paper (7.5 cm x 15 cm [3″ x 6″])
  Newspapers
  Children's scissors
  Plastic cup
  Clear, plastic containers (wide variety: small and large, wide and narrow)
  Glass containers (wide variety: tall vinegar bottles, fat pickle jars, small relish jars)
Preparation/Demonstration:
  Scissors

HOW we will do it: Spread newspaper on the activity table and mix up the paints. Cut out the 7.5 cm x 15 cm (3″ x 6″) pieces of paper, and fold them in half.

Ahead of time, make a sample picture by finger painting on one side of the paper, folding the blank side onto it, and pressing down. Cut the paper apart in the middle after the paper has dried. You have now made two identical pictures. When one is placed behind or underneath a magnifier, you will be able to compare them and to observe how much magnification is taking place.

Push a long narrow table against a wall. Fill the plastic cup with water and set it on the table. Tape one of your identical pictures on the wall near the cup, but not directly behind it.

During your Attention Getter time, show the children the picture on the wall, near the cup of water. Show them the second, identical picture. Ask your students to compare the two pictures, noticing in particular their size and shape. Next, ask the children to predict if the second picture will still look the same when it is put behind the cup of water. Tape it there, and ask the children what they notice. How does the image look different? Why? Show your students all the other jars, bottles, and containers. Ask them to predict whether these vessels alter the image when filled with water. Suggest to the children that they make their own identical pictures to use in conducting the experiment, and ask them if they see anything in the room which will help them make their pictures.

Bring out a folded piece of blank paper, and show the children how they can paint one side, bring the other side over, and press it down on the wet paint. Use the word *identical* when you talk about the pictures. Let the children know that their identical pictures will be different from yours and everyone else's because each person has his or her own way of making things. Put your sample away before the children begin to explore the materials. When the paintings are dry, cut them along the crease lines in the middle.

Water Magnifying Activity (Part 2)
Science

WHY we are doing this project: to enable children to experiment with the magnifying ability of water.

The facts of the matter: Light that enters a glass or jar of water is bent in the same way as light that enters a magnifying glass and, therefore, magnifies.

WHAT we will need:
  Glass and plastic containers (collected for Part 1)
  Masking tape
  Identical pictures (created in Part 1)
  Water
  Small pitchers
  Duct tape
  Long, narrow table
Optional:
  Cautionary sign (provided on page 7; photocopy and enlarge for your use)
  Puppet

HOW we will do it: Using glass jars and bottles is optional, however they greatly enhance this experiment. Even if you work with very young children, you can discuss ahead of time the rule that the glass jars and bottles are not to be touched or moved. Make sure all your tall, narrow containers are plastic, and if you like, demonstrate the magnifying ability of tall, narrow glass bottles yourself during an Attention Getter time. Talk about what would happen if a glass jar or bottle was dropped. Pin up the cautionary sign beside the materials and discuss it. You know best what the listening abilities and energy levels of your children are, so use only plastic containers if you feel it is necessary—the experiment will still be an interesting one.

Move the long, narrow table up against the wall and line up your containers on it. Tape half of each identical picture made in the previous activity behind the container and jar magnifiers. Make sure you have enough jars and containers to be able to tape every child's picture to the wall.

Provide the children with small pitchers of water and ask them to fill the containers. Invite your students to find their pictures on the wall, and to tape the matching picture on the table, in front of the magnifier behind which the identical picture is placed. Ask them to compare the two identical pictures. Do the magnifiers change the way the pictures look? How?

Encourage examination and comparison of the different magnifiers. Take out your puppet to ask the following questions, or ask them yourself: What do they notice about which container is the best magnifier? What do they notice about the difference between how plastic

Please don't move the glass jars!

containers magnify and how glass containers magnify? Set out an empty container on your table also. What do the children notice when a picture is placed behind it? When experimenting with tall, short, wide, and narrow containers, what do the children notice about which ones act as better magnifiers? Encourage observations and comparisons. Explain that light bends when it passes through the water, and bends again when it leaves. This makes the pictures behind the jars of water look bigger.

Tip: Leave the containers in this arrangement for the next activity.

Another Magnifier Tester
Art/Math/Science

WHY we are doing this project: to encourage creative expression; to provide experience with creating patterns; to develop fine motor skills.

WHAT we will need:
    Inexpensive, plastic rulers (one for each child)
    Stickers (many of the same shape and color)
    Spangles (many of the same shape and color)
    Glue
    Paper
    Markers
    Containers filled with water (from previous activity—omit any tall, narrow, glass bottles)

HOW we will do it: In this project, children will design a long strip of patterned art to glue onto their rulers. When they pass the ruler behind a magnifier, they will be able to compare the magnified image with the same non-magnified image on the rest of the ruler. This will allow them to see just how much magnification the bending light in each container creates. For this experiment, make sure all the glass jars are large with wide circumferences so that they are difficult to move when filled with water. Secure them to the table with duct tape. Point again to the "Please don't move the glass jars" sign and discuss it with the children.

To prepare, cut out strips of paper that are the same size as the rulers. Set out all materials on the activity table. Ahead of time, make a sample by gluing a strip of paper onto a ruler, and then making a pattern along the strip, using glue, spangles, stickers, or your own drawings. Remove all tall, narrow, glass vessels from your collection.

During an Attention Getter time, show the children your magnifier tester and tell them that that is what it is called. Point to each shape in your pattern and encourage the group to help you identify out loud what each one is. For example (pointing to each shape in the pattern): "First there's a red heart, then a yellow star, then a red heart, then a yellow star, then a _____." After a while, stop speaking yourself and let the children speak alone. Tell the children that when something has the same shapes or colors in the same order, over and over again, it is called a *pattern*. Ask the children to predict what they will see when they put part of a magnifier tester behind each container. Ask the children what they see in the room that will help them make their own magnifier testers. (Remember to mention that everyone's pattern will be different because everyone has their own special way of doing things. Put your sample tester away before the children approach the materials.)

If children do not make patterns on their testers, let that be okay. Younger children will probably choose shapes at random for their testers, but will still see magnification when they pass them behind the containers. For older students, you can put a more complex pattern on your sample tester to challenge them; for example, make a pattern using three shapes or four. For kindergartners, you might choose to make a sample pattern that is the same in shape but not in color, or that has the same color order but not shape.

After the children make their testers and the glue has dried, encourage them to place part of their testers behind each container. Ask them to compare the size of the images that are behind the container with the images that are not. Do they notice whether any of the shapes are changed in other ways, besides being magnified (for example, twisted or distorted)? Encourage the children to express their findings.

Art kits: In large-sized sandwich bags, put a ruler, spangles, sequins, stickers, a few crayons and markers, a glue stick, and a strip of paper that is the same size as the ruler. Put these magnifier-tester art kits out for the children to use as and when they choose.

Fluids As Magnifiers
Science

WHY we are doing this project: to help children experiment with how different fluids magnify; to compare their relative magnifying power; to facilitate fine motor development through pouring.

WHAT we will need:
Trays
Identical, small, clear, plastic containers (clear, plastic cups work well)
Funnels
Four small pitchers
Vinegar
Oil
Water
Corn syrup
Magnifier testers (created in previous activity)
Prediction chart (format provided on page 73; photocopy and enlarge for your use)
Butcher paper
Markers
Newspapers

HOW we will do it: To prepare, tape up the piece of butcher paper near the place where you gather for Attention Getter time. Use the format provided to make a prediction chart. Line up the containers on the trays. Pour a small amount of each liquid in its own pitcher.

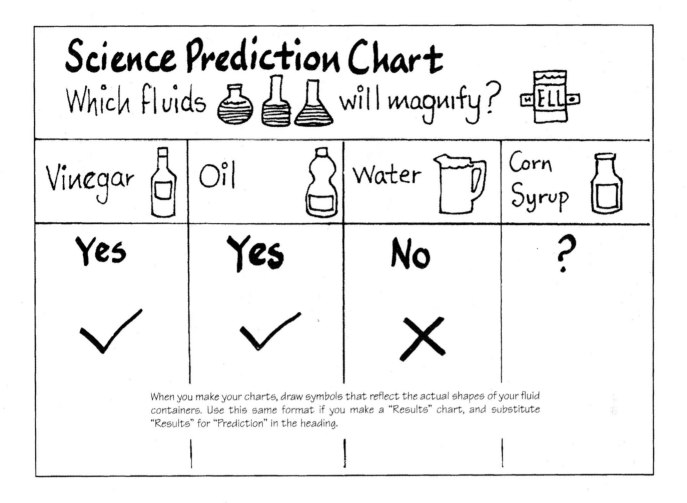

Science Prediction Chart
Which fluids 🧪🧪🧪 will magnify?

Vinegar	Oil	Water	Corn Syrup
Yes	Yes	No	?
✓	✓	✗	

When you make your charts, draw symbols that reflect the actual shapes of your fluid containers. Use this same format if you make a "Results" chart, and substitute "Results" for "Prediction" in the heading.

Set everything on the activity table. Be sure to spread plenty of newspaper underneath, as the children will be pouring from the pitchers to the containers.

During an Attention Getter time, take out your magnifier tester, and tell the children that so far you have only used it with a magnifier made out of a container of water. Say: "You know that water is a magnifier, but do you think other liquids are too?" Show the children the original bottles that contained the vinegar, oil, and corn syrup and, together, identify the fluid in each container. Show the children the corresponding symbol for each fluid on the prediction chart. Let your students smell the liquids if they would like to.

Ask them to predict whether or not each one will magnify. Print, or have the children write, a "yes" and a check mark for each prediction that a fluid will magnify, and a "no" and an "X" mark for each prediction that it will not. Print the childrens' names next to their pre-

dictions or have your students do this. Show them the blank spaces on the chart, and let them know that after they conduct their experiments, they can write yes or no in these spaces. Be prepared to let younger children scribble on the chart.

After the children have poured their liquids, passed their magnifier testers behind the containers, and generally explored to their hearts' content, gather together at the end of the day or session and compare their predictions to the results. If you like, make another Science Results Chart, and record your discoveries on it. Then compare the two charts.

Making Water-Drop Magnifiers

Science/Fine Motor

WHY we are doing this project: to help children understand that drops of water are convex lenses; to help them discover that water drops bend light in the same way that magnifying glasses do and that therefore they also magnify; to develop fine motor skills.

WHAT we will need:

Plastic cling wrap
Cardboard
Crayons
Markers
Staplers
Eyedroppers
Small containers
Water
Photocopies of design page (provided below) or magazine pictures

Preparation:

Scissors
Glue stick
Paper

HOW we will do it: As this activity requires a lot of preparation, ask friends, neighbors, and other parents to help you with this ahead of time. To prepare, cut out cardboard squares measuring about 7.5 cm x 7.5 cm (3" x 3"). Fold each one in half, and cut out a rectangular hole out of the fold, about 13 mm (½") from the edges, so that when you unfold the cardboard, you have a frame. Make two of these for each child. Cut out photographs from magazines, especially tiny letters or pictures, and/or make photocopies of the design page, and cut and paste some of the tiny pictures onto paper. Cut out cling wrap squares which also measure about 7.5 cm x 7.5 cm (3" x 3"). To store them in preparation for the activity, it works best to lay them very loosely in a box with a lid.

Put small amounts of water in the containers, and set all materials on the activity table. Ahead of time, make a sample water magnifier. Stretch a square of cling wrap between two frames and staple around the edge. Using crayons and/or markers, decorate the frame of your magnifier. Have one of the papers with tiny pictures available.

During an Attention Getter time, ask the children what they remember about the magnification properties of water in plastic and glass containers. Ask if they remember what light does as it travels through water. (It bends.) Show the children your water magnifier and say: "This is a tool I made to test whether one drop of water will magnify." Explain how you made it, and indicate the appropriate materials on the activity table.

Use an eyedropper to drop a small bead of water onto the cling wrap. Show your students one of the miniature pictures, and without demonstrating this, ask them to predict whether the tiny bead of water will magnify the picture. Ask the children how they could conduct this experiment for themselves and after listening to their responses, invite them to use the materials on the activity table. You will probably have to help with stapling.

Encourage the children to examine the tiny pictures with their water magnifiers. What do the children notice when a picture is examined through the bead of water? What does this tell us about whether or not a single drop of water will magnify? If necessary, mention again that as light goes in the drop of water and out again, it is bent as it passes through.

What Can We See with a Magnifying Glass? (Part 1)

Science/Language

WHY we are doing this project: to enable children to watch a stain spreading with the aid of a magnifying glass; to give children an appreciation for the benefits of using a magnifier; to develop speaking and listening skills.

WHAT we will need:
Magnifying glasses
Medicine or eyedroppers
Small containers
Water
Fabric scraps
Mustard
Ketchup
Popsicle sticks
Styrofoam meat trays
Puppet

HOW we will do it: To prepare, set your meat trays out on the activity table. On each one, place an eye- or medicine dropper, a magnifier, and a fabric scrap. In the middle of the table, put out small containers of water, ketchup, and mustard. Put popsicle sticks into the mustard and ketchup containers so that the children can smear dabs of them onto the fabric. The containers should be arranged so that all children have access to them.

During an Attention Getter time, point out the materials to the children, and invite them to use the droppers to let small globs of mustard or ketchup fall on their fabric. Encourage them to use the magnifying glasses to watch how the stain spreads. What do they notice about whether the fabric absorbs the mustard or ketchup quickly or slowly? What happens if the children let drops of water fall onto the fabric first? What do the children notice about whether this changes how the stains spread? As the children conduct this experiment, take out your puppet and use its personality to ask your students what they are doing. Ask them to look at a spreading stain without a magnifying glass, and then with one. Which method allows them to see what is happening most clearly?

What Can We See with a Magnifying Glass? (Part 2)

Science/Language

WHY we are doing this project: to give children the opportunity to observe a chemical process with the aid of a magnifying glass; to further develop an appreciation for what magnifying glasses can do; to develop speaking and listening skills.

WHAT we will need:
In addition to the materials from the previous activity:

Liquid dishwashing detergent
Liquid starch
Bar of soap
Preparation:
Grater

HOW we will do it: To prepare, grate a small amount of soap from a soap bar. After the children have used their magnifiers to watch stains spread, put out small containers of liquid starch, liquid dishwashing detergent, and grated soap. Encourage the children to watch what these soaps do to their stains under a magnifying glass. Does one type of soap work on the stains more quickly than others? What happens when the children add drops of water on top of the soap and/or on top of the stain? Use your puppet's personality to talk to the children about their experiment. Ask them how the magnifying glasses affect what they see.

Paper Textures and Magnifying Glasses
Science/Sensory

WHY we are doing this project: to expose children to different paper textures through the use of magnifiers and to provide a sensory experience.

WHAT we will need:
Embossed paper
Business cards with raised print
Aluminum foil candy wrappers
Sandpaper
Newsprint pieces
Any other interesting textured paper

HOW we will do it: To collect your paper samples, contact business offices, office supply stores, paper supply stores or photocopy shops to ask for donations of samples. Explain that

you want the materials for children. You can also ask friends, family, and other parents to help you collect these items.

On the day of the project, lay out magnifying glasses and samples of the different papers on an activity table. Encourage the children to explore, and invite them to express what they observe. Invite them also to touch the different papers with their fingertips. How do they feel different?

Science kits: Collect some small containers and/or boxes with lids. Put a magnifier and a collection of different, textured paper squares in each one. Set these out for the children to use as and when they like.

Examining Other Materials with Magnifiers
Science

WHY we are doing this activity: to enable children to closely examine a variety of everyday materials with magnifying glasses.

WHAT we will need:
Magnifying glasses
Leaves
Stamps
Alphabet noodles
Coins
Small shells
Small pebbles
Salt
Pepper
Preparation:
Poster board
Glue stick

HOW we will do it: These items are only suggestions. If you have other interesting materials you would like the children to examine in this activity, then by all means, use them. To prepare, cut out small poster board squares, and glue some salt, pepper, a stamp, and some alphabet noodles on each one. If you do not have time to do this, you can put some of each of the materials in a small plastic or Styrofoam tray. Set these and the other materials out with the magnifying glasses, and encourage the children to explore them and to verbalize their observations.

Science kits: Place a variety of these materials in several containers with lids and put a magnifier in each one so that children can engage in independent exploration.

Measuring How Much Magnifiers Magnify
Science/Math

WHY we are doing this project: to enable children to measure and compare how much a magnifying glass enlarges objects.

WHAT we will need:
    Magnifying glasses (preferably of
       varying strengths)
    Graph grids (provided below;
       photocopy and enlarge for your use)

HOW we will do it: By using the uniform grids of the graph papers, the children can compare magnified graph lines with graph lines which are not magnified. This will give them a more exact idea of just how much a magnifier enlarges

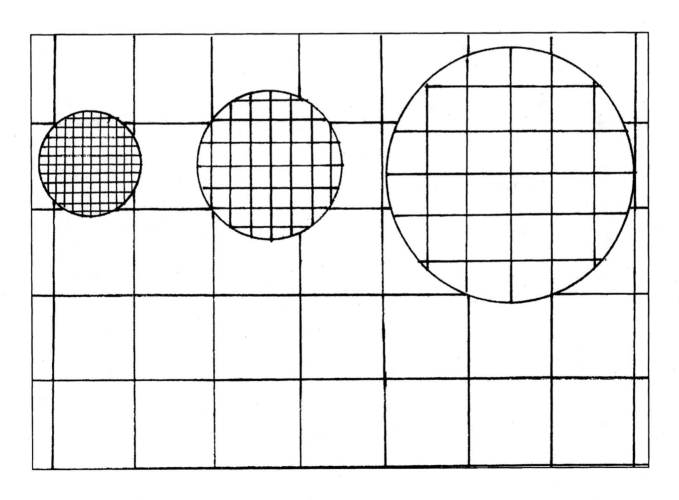

an image. If possible, invest in at least one high-quality magnifying glass. A four-power magnifier is of high quality and ranges in price from $5 to $13. Hand-held sixteen-power lenses are also available for between $10 and $15. You can find magnifying glasses in the optics section of nature stores.

Ask your students to predict how well each tool will magnify. As the children participate in this activity, encourage them to compare how much the different magnifying glasses enlarge the graph squares, and have some one- and two-power (less expensive, lower quality) magnifiers available to provide a contrast. Encourage your students to verbalize their findings as they experiment.

Magnifying Books
Science/Language

WHY we are doing this project: to develop self-esteem and initiative by encouraging children to find and choose their own materials for examining under magnifiers; to develop self-esteem by giving the children their own magnifying glasses; to develop fine motor skills; to develop all components of language arts: reading, writing, speaking, and listening.

WHAT we will need:
    Construction paper
    Glitter or colored sand
    Spangles
    Pepper
    Salt
    Magazines
    Leaves
    Fabric scraps
    Yarn
    Glue sticks
    Small, plastic magnifying glasses
    Crayons
    Markers
    Pencils
    Pens
    Scissors
Preparation:
    Hole puncher

HOW we will do it: In this project, you will provide the children with blank books and small, plastic magnifying glasses. The children will search for, choose, and glue in objects for examination with their magnifiers, and then add words about each one.

If you work with a large number of children, you can order inexpensive, small magnifying glasses from school supply catalogs. Also, they are sometimes available at toy or nature stores.

To prepare, cut the construction paper into 25 cm x 15 cm (10" x 6") pieces. Take three sheets and fold over the 25 cm (10") side. Then staple along the fold. Make one blank book for each child. Use the hole puncher to punch a hole in the upper left corner, tie yarn around the hole, and tie the other end around a magnifying glass. Make sure the yarn lengths are long enough for children to freely move the magnifier to examine objects on each page of the book.

Make a sample magnifying book. Several weeks ahead of time, begin collecting tiny and large objects which can be examined closely with a magnifier. Glue an object or objects onto each page, and write a few words about each. For example: "I found this little piece of foam on the floor, in a corner. When I look at it through my magnifying glass, I can see little holes in it." On the cover of your book, write a title and "by (your name)." Leaf through your magazines, and cut out some of the pictures you find. Also, tear out pages to put on the activity table for children who want to do their own cutting.

On the day of your activity, set out spangles, glitter or colored sand, salt, pepper, and the pictures. Cut the leaves, fabric, and yarn into pieces and set these out as well. Put out any other safe objects that could be glued into the books. Set out the blank books and glue sticks. This will be an ongoing project that begins with this activity. On a small table, make a writing center for the books the children have begun. Set out the glue sticks, scissors, and writing and drawing materials so that, as the children find materials for their books at home or in other places, they can continue to add to their magnifying books.

During an Attention Getter time, show the children the sample magnifying book and read it to them. If you like, pass it around so that the children can examine the objects in it with the attached magnifying glass. Tell the children about places you looked to find things for your book (e.g., corners, outside, in closets, under beds). Ask the children what they see in the room that would help them make their own magnifying books.

As they work on the books, take story dictation, spell words, support invented spelling or scribbling, or write words to be copied onto the book pages, depending on the needs of your students. As the children finish the project, encourage them to look in other places for things to add to their books. Together, brainstorm for places they could search. Talk about safe and unsafe places to look, as well as safe and unsafe things to touch or pick up.

Leave the books and other writing center materials on the small table for several days. If you like, have a show-and-tell of magnifying books several days later. This does not have to be formal; you can simply make a point, during the day or session, of asking each of your students to read their books to you, or if they prefer, reading the books to them.

Magnifying Boxes
Math

WHY we are doing this project: to practice rational counting; to develop reading skills; to further develop an appreciation for magnifying glasses.

WHAT we will need:
    Magnifying glasses
    Trays
Preparation:
    Small, cardboard jewelry boxes
    Glue stick
    Glitter or colored sand
    Spangles
    Small seeds (for small birds like finches)
    Small beans
    Other tiny objects
    Permanent marker

HOW we will do it: In this project, you will need to glue a small number of tiny objects inside a box. The children will then count them, using a magnifying glass. Begin collecting your jewelry boxes ahead of time.

In small, cardboard jewelry boxes, use a glue stick to smear some glue on the inside and then drop some single seeds, spangles, glitter, and so forth onto the glue. It is important that each tiny object have space around it to facilitate counting and examining. An easy way to do this is to press your finger down on the object and then let it drop onto the glue. If you work with kindergartners, drop fifteen to twenty objects in each box. If you work with younger children, drop fewer than ten in each box. On the lid of every box write: "How many?" with a permanent marker.

Put your object containers and magnifying glasses on trays and encourage the children to use the magnifiers to examine the tiny objects and to count them.

Developmental differences: Three- and young four-year-olds will be most interested in opening and closing the containers and looking at the things inside, with or without magnifiers. If you like, ask the children what they see in the boxes, and whether they would like to help you count them. Older children will be more likely to count the objects by themselves.

The Mystery of the Missing Banana Cream Pie
Language/Multicultural/Anti-Bias

WHY we are doing this activity: to develop speaking and listening skills; to present children with a different perspective on gender roles; to develop multicultural awareness.

WHAT we will need:
    Flannel board
    Flannel board shapes (provided on
        page 8; photocopy and enlarge for
        your use)

Felt
Glue
Scissors
Markers or crayons
Optional:
Clear contact paper
Story: "THE MYSTERY OF THE
MISSING BANANA CREAM PIE"
(provided)

HOW we will do it: If you can, get help coloring the flannel board shapes from friends, family, or other parents. Glue the colored pieces onto felt and cut them out. If you want to make them last longer, cover the paper side with clear contact paper. Set up your flannel board and tell this story:

"THE MYSTERY OF THE MISSING BANANA CREAM PIE"

One day a little girl named Charlotte Holmes [put Charlotte on flannel board] made a beautiful banana cream pie and set it on the kitchen table, ready to eat. [Put table with pie on board.] All of a sudden the telephone rang, and Charlotte went to answer it. [Take Charlotte off board. Take table with pie off board and replace with the table and empty pie dish.] But when Charlotte got back to the kitchen, guess what she found? [Let children answer: the pie was gone.]

"Where's my pie?" Charlotte yelled. "It took me hours to make it!! Where is it? Who took it? I can't believe this!" Charlotte was very upset. She looked under the table and on the floor but there wasn't a trace of that wonderful banana cream pie. "Hmmm," Charlotte said. "This is a real mystery. I'm going to have to use my detective tools to get to the bottom of this." [Take Charlotte off board; replace with Detective Charlotte.] A minute later Charlotte had everything she needed to unravel the mystery. [Ask children: What is Charlotte wearing on her head? (Hat) What is she wearing around her shoulders? (Cape) What is she holding in her hand? (Magnifying Glass).]

"Okay, Banana Cream Pie Thief, whoever you are, look out for Charlotte Holmes, Detective Extraordinaire," Charlotte said. The first thing Charlotte did was to examine the pie plate under the magnifying glass, and she noticed that it looked as if a very large tongue had been licking it. Then the magnifying glass showed her a big white whisker in the pie dish. Do you think that whisker belonged to Charlotte? [Let children answer.] Then she looked at the floor again with her magnifying glass, and guess what she saw this time? [Put up paw prints.] What do you think made those prints? [Let children answer.] "Hmmm," Charlotte Holmes said. "The licked pie dish, the big, white whisker, and these paw prints are all very important clues to solving this mystery."

Using her magnifier, Charlotte Holmes followed those paw prints and guess where they lead? [Put up dog in dog basket.] She found Ralph lying in his dog basket, and guess what she found all over his mouth? [Let children answer.] Ralph was so full of banana cream pie that he was lying on his side, moaning and groaning. How do you think Charlotte felt? [Discuss: Angry at Ralph? Sorry for Ralph? Disappointed that she couldn't eat the pie? All of these things?] Well, Charlotte scolded Ralph for stealing that banana cream pie, and then she wrapped him up in a blanket and let him sleep. The next morning Ralph felt better, and Charlotte made another banana cream pie. Do you think she left it on the kitchen table this time? [Let children answer.]

After telling this flannel board story ask the children for stories about their pets. Set the materials out for the children and encourage them to use them to retell the original story and to make up new ones. Make a "Two people may be here" sign and pin it up near the flannel board.

VOCABULARY:
Trace
Upset
Get to the bottom of
Mystery
Unravel
Detective
Extraordinaire
Scolded

Detective Play
Dramatic Play/Language

WHY we are doing this activity: to enable children to engage in fantasy play; to encourage child-to-child interactions; to develop language; to develop an enjoyment of acting.

WHAT we will need:
Dress-up raincoats
Dress-up hats
Plastic magnifying glasses
Notebooks
Pens

HOW we will do it: Several weeks ahead of time, start asking friends, families, and other parents if you can borrow hats and raincoats for this activity. Set out all materials in a dramatic play area. Tell the flannel board story above, or read some of the detective stories listed in the literature section. Afterward, show the children the capes, hats, and magnifying glasses and invite them to create and solve their own mysteries.

Magnifier Song
Gross/Fine Motor/Music/Movement

WHY we are doing this activity: to help children feel comfortable using their singing voices; to reinforce the function of magnifiers through music; to develop the large and small muscle groups.

WHAT we will need:
Song: "MAGNIFIER SONG" (to the tune of "Hello Operator")

"MAGNIFIER SONG"
Had a diamond in my pocket;
the pocket had a hole,
the diamond worked its way right out,
and then away it rolled.

I got my magnifier,
I scrutinized the floor,
I saw my diamond clear as day
fetched up beside the door!

HOW we will do it: Sing the song once for the children and explain the words *scrutinized* and *fetched up*. For "Had a diamond in my pocket," pretend to hold a glittering diamond between your thumb and forefinger, and put it in your pocket. For "the pocket had a hole, the diamond worked its way right out," pretend to pull out an empty pocket and make a dismayed face. For "and then away it rolled," roll your hands around each other. For "I got my magnifier," pretend to hold a magnifying glass and stand up. For "I scrutinized the floor," pretend to examine the floor through your magnifier. For "I saw my diamond clear as day, fetched up beside the door," pretend to find your diamond (by a door if possible) and pick it up with delight.

Sing the song and perform the motions together as a group. If you have an adequate number of plastic magnifying glasses, use these during the song.

Literature

Symbol Key: *Multicultural
              +Minimal diversity
              No symbol: no diversity or no people

The first three books all have several illustrations of children (or bears!) using magnifying glasses.

Berenstain, S., & Berenstain, J. (1980). *The Berenstain bears and the missing dinosaur bone*. New York: Random House.

Levy, E. (1975). *Something queer at the ball park: A mystery*. New York: Delacorte Press.

Levy, E. (1975). *Something queer on vacation: A mystery*. New York: Delacorte Press.

Norden, R. (1993). *Magnification*. New York: Lodestar. (This is an excellent pop-up book with wonderful photographs of various magnified objects.)

Extenders

Science: Encourage the children to lie on the floor and examine it with their magnifying glasses. Can they spot things they would not ordinarily see? What do they find in corners and against the walls? Invite the children to examine other objects in the room with magnifiers, for example, walls, carpet, wooden furniture, dust balls.

Science: Provide cooked and uncooked alphabet noodles for the children to examine and compare. How do they look different under the magnifying glasses?

Cognitive: Buy very small stickers and/or use the tiny pictures provided in the book. Spread butcher paper out on a table and tape it down. Affix the tiny pictures or stickers all over the butcher paper, and put matching pictures/stickers on separate index cards. Invite the children to pick a card, and to use their magnifiers to see if they can find the matching sticker or picture on the butcher paper. Make sure you pull the table away from the wall so that the children can move freely all around it. A long, narrow table is preferable to a square or circular one.

Music/Movement: As a variation of the "MAGNIFIER SONG," use a plastic bead which looks like a diamond. (These are available from craft stores.) Have one child hide it while everyone closes their eyes. Change the last line to: "I couldn't ask for more!" As you sing the song and perform the motions, see if anyone can find the hidden diamond.

Math/Art: Let the children examine fabric pieces under a magnifying glass. Ask them what they notice about the threads, and how the individual threads are arranged to make the fabric. Introduce the word *weaving*, and then provide the children with strips of paper to weave in and out of each other.

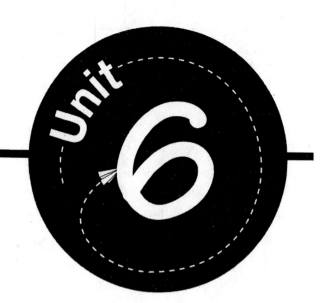

OUTER SPACE

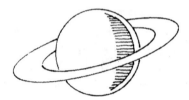

Facilitating a unit on outer space is a challenge, because it is hard to keep activities hands-on and child-initiated when the unit focuses on something—the solar system—that children cannot touch, visit, see, or experience directly. This unit, like the others, is designed to be as hands-on as possible, and I have to say my students have always enjoyed it tremendously.

One thing that helps immensely is to create a "space environment" in your home or classroom, as much as is feasible. Before you begin the unit, put space posters up on your walls and play "spacey" music. Be creative!

Another important aspect of this unit is to help children feel that everyone has an equal opportunity to choose a career as an astronaut or in space technology. This can be very difficult to do since there are few photographs available of astronauts and space scientists who are not white men. For this reason, I have included a clip art section in this book which provides pictures you can use to enlarge by photocopier and post on your walls to create a multicultural, anti-bias environment for this unit. Cut out diverse faces from magazines to glue into the astronaut helmets. Another way you can combat societal bias during your pretend spaceship journeys is to give every child a turn as Mission Leader.

Attention Getter: See the list of books on outer space at the end of this unit, and check out a few of them, or others, from the library. Try to find books with actual photographs of outer space (or at least very realistic drawings). Leaf through the books with the children, and discuss what you see. Identify planets, stars, the moon, planet Earth, satellites, and so forth. Ask the children to guess what you will be talking about and working with over the next few days.

NASA Outer Space Kits
Science/Multicultural/Anti-Bias

WHY we are doing this project: to obtain excellent posters, pictures, booklets, and information about outer space and space technology; to help all children feel that a career in space science is open to them.

WHAT we will need:
 This address:
 NASA Publications
 250 E Street, Southwest
 Washington, DC 20024
 1-202-358-1110

HOW we will do it: To make sure you get your NASA kit in time, write or call about one month before facilitating the unit. Be sure to specify that you want the kit designed for the lowest grades (preschool to kindergarten).

I do not know how frequently NASA develops new materials or changes the contents of the kit, but you may want to specifically request the materials below.

The "NASA Superstars of Science" poster portrays and describes the work of scientists from diverse races and backgrounds. Another wonderful poster (although I could do without the Native girl's headband) is titled, "There is a new world for all of us," and shows children of different races in astronaut suits, floating above the earth. Also excellent is a poster called "Spacesuit," which names and describes the function of every component of a space suit. (A booklet is available which elaborates on this.) "Rockets" and "Aboard the Space Shuttle" were also included in the kit I received, as well as some fantastic 20 cm x 25 cm (8" x 10") photographs of planets, rockets, and astronauts. An "Information Summaries Source List" provides addresses for other space-related organizations. The kit also included books with activities, although many of these are not developmentally appropriate for young children. However, if you need some excellent posters for the walls while you do this unit, as well as some good information, then order a kit. Our tax dollars at work—the kit is free!

Earth Is Not Alone
Science/Small Group Activity

WHY we are doing this project: to familiarize children with the planets; to help them remember the planets' names; to help them understand the size of Earth in relation to the other planets; to familiarize children with ordinal numbers; to help them remember that our sun and the nine planets which revolve around it comprise the solar system.

WHAT we will need:
 Poster of the solar system
 Poem: "THE PLANET POEM"

"THE PLANET POEM
Mercury's closest to the sun,
Pluto is the farthest one.
From the sun, just where is Earth?
When you count them, it is third.

HOW we will do it: In poster shops, it is usually easy to find some excellent charts that show the planets of the solar system. Try to find one that uses bold colors, aims for realism in terms of the colors of the planets, and that, ideally, has fairly large type to identify each one. There are two layouts: one shows the planets in a line from the sun. This is the kind you need for this activity so that the children can clearly see that Earth is third closest to the sun. However, I also recommend that you buy a poster that shows the planets around the sun, so the children understand that this is how the planets are positioned, and are not in a straight line.

Pin up the poster on the wall and discuss it during an Attention Getter time. Let the children know that, together, the sun and nine planets are called the *solar system*. *Sol* is the Latin word for sun. Explain that even though the poster makes the planets look very close together, they are actually millions and millions of miles apart. From the sun, the planets are: Mercury, Venus, Earth, Mars, (Asteroids,) Jupiter, Saturn, Uranus, Neptune, and Pluto. Say the above rhyme several times until the children know the words. Together, count the planets on the poster, starting with Mercury, which is closest to the sun. Which one is third? Compare its size to the other planets. What do the children notice about whether it is the biggest? What do they notice about whether it is the smallest? Use the poster as a backdrop for the following activities.

Making Space Helmets
Art/Fine Motor

WHY we are doing this project: to prepare for the dramatic play activity to follow; to facilitate creative expression; to stimulate interest in the unit.

WHAT we will need:
    Large, brown paper grocery bags
    Aquarium tubing (3 mm–6 mm [⅛"-¼"]
        in diameter)
    Crayons
    Markers
    Poster (used in previous activity)
    Picture of astronauts (provided on
        page 87; photocopy and enlarge
        for your use)
Preparation:
    Scissors
    Cardboard
    Glue
    Exacto knife
    Wooden cutting board

HOW we will do it: To prepare, cut your grocery bags down so that children can wear them on their heads. Cut a window out of each one.

In the following activity, the children will make "air packs," which will then connect to the space helmets with aquarium tubing. You will need to create a stiff opening in each helmet into which the tubing can be inserted. To do this, cut out cardboard squares about 7.5 cm x 7.5 cm (3" x 3"). Glue one cardboard square securely to each helmet, at the front and near the bottom.

After the glue has dried, lay one of the bags down on the wooden cutting board, with the cardboard square facing out.

In the center of the square, from the inside of the paper bag, cut a cross (+) in the center, leaving a wide margin of cardboard around the cross. Use an Exacto knife or razor blade to do this. When the tubing is inserted later, the cross will hold it securely in place. Prepare enough helmets for all the children, as well as one sample for yourself.

During an Attention Getter time, tell the children that they are going to take a trip to outer space, and that they will have to start preparing for it. Point to the poster of the planets, and say, "Do you think these planets have air around them, like the air we breathe?" Explain that the other planets have gases around them that human beings cannot breathe. Show the children the picture in the text of the astronauts, and notice how they are all wearing space helmets. Explain that the hel-

Photocopy these pictures on a photocopier

mets protect the astronauts from breathing harmful gases.

Put your own "space helmet" on, and tell the children that you made it, and that, when you wear it, you pretend that the window is covered with plastic, like astronauts' real helmets. Set out the prepared bags, crayons, and markers and invite the children to decorate their own space helmets.

Making Life-Support Systems
Art/Fine Motor

WHY we are doing this activity: to stimulate interest in the unit; to develop self-esteem and a sense of autonomy by enabling children to make their own props for dramatic play; to prepare children for the Traveling in Outer Space activity.

WHAT we will need:
Plastic gallon jugs (two for each child)
Aquarium tubing (from previous activity)
Control panel illustrations (provided on page 88; photocopy and enlarge for your use)
Construction paper
Glue
Small containers for glue
Glue brushes
Crayons
Masking tape
Children's scissors
Preparation:
Scissors
Rope (several pieces, 1'8" each)
Duct tape

HOW we will do it: To prepare one "life-support system," glue two gallon jugs together. The handles should extend out to the left and right of the pair. Even after the glue dries, it is easy to pull the jugs apart again. Wind duct tape around the two containers to attach them more securely.

Loop a separate length of rope through each handle, and tie the ends together. The children will put their arms through the loops so that the air packs hang on their backs like backpacks.

Children enjoy having the life-support systems as props for this unit; it really adds a dimension to the dramatic play, and to the Traveling in Outer Space activity below. However, there is quite a bit of preparation involved. One alternative is to prepare three or four packs and have the children decorate them as a group project. Let the children take turns using them during dramatic play and omit them from the following activity. If you do want all the children to have the opportunity to make their own, enlist friends, family, or parents to help you with the preparation.

On the activity table, set out the glue, glue brushes, prepared air packs, the copies of air control panels, construction paper, children's scissors, and crayons.

During an Attention Getter time, point to the life-support system on the NASA poster of an astronaut's equipment. Ask the children what they think it is for. Discuss with the children the other parts of an astronaut's life-support system. Connect a sample helmet and life-support system with the aquarium tubing and put them on. Explain to your students that if you are going to explore space, you will need them both. Ask them if they see anything in the room that they can use to make their own life-support systems. Encourage them to be creative with the markers and crayons. Help them to secure one end of the tubing to the jugs with masking tape and to poke the other end through the cross in their space helmets.

Making Space Glasses
Art/Crafts

WHY we are doing this activity: to facilitate artistic expression; to develop fine motor skills; to help children understand that the sun is so bright it can damage our eyes (therefore, we must never look at it directly).

WHAT we will need:
- Cardboard
- Pattern for space glasses (provided on page 90; photocopy and enlarge for your use)
- Acetate papers (different colors)
- Thick rubber bands
- Glue
- Small containers for glue
- Glue brushes
- Space stickers
- Crayons
- Markers
- Aluminum foil scraps

Preparation:
- Scissors and/or Exacto knife
- Stapler

HOW we will do it: Use the pattern provided to cut the frames for space glasses out of cardboard. Ask one of the children to model for you, so that you can determine whether you need to make the frames bigger or smaller. You will be stapling the ends of cut rubber bands to the sides of the glasses, so keep in mind that the elastic must stretch a little to secure the glasses on each child's head.

Use the scissors and/or Exacto knife to cut a window out of the glasses, as shown in the pattern. Cut out rectangles of acetate paper, just slightly smaller than the glasses. Glue the acetate to the back of the glasses. Trim off the excess acetate that extends below the curve of the glasses, where they rest on the children's noses. Cut each rubber band once. Staple one end of the band to each side of the glasses, as indicated on the pattern. Prepare enough glasses for all the children, as well as a sample pair for yourself.

Put all the materials on the activity table so that the children have access to them.

During an Attention Getter time, show the children a picture of the sun from one of the outer space books. Ask: "What do you think would happen if you looked right at the sun for a few seconds? (You can damage your eyes.) Say: "We're going to visit the sun on the space trip we're going to be taking, so we will need space glasses to protect our eyes." Put on your own glasses. Ask the children if they see anything in the room that they could use to make space glasses. Invite them to be creative with the stickers, crayons, markers, foil, and glue.

Traveling in Outer Space
Science/Dramatic Play/Movement/Small Group Project

WHY we are doing this activity: to help children learn how the planets are different and to stimulate interest in the unit.

To keep your curriculum child-initiated, give children a choice about participating in this activity. If some would rather not, provide a space with toys and projects for them to use while you do this activity with those who want to join you.

WHAT we will need:
- Space helmets, space glasses, and life-support systems (from previous activities)
- Long rope
- Chairs
- Space poster
- Calculators
- Flashlights
- One orange
- Long skewer
- Paper bag

HOW we will do it: Arrange the chairs to simulate a spaceship. Skewer the orange and hide it in a paper bag. Tell the children that you are going to take a trip to all the planets, so you will need your space helmets, space glasses, and life-support systems. Encourage your students to put on their equipment, and put a helmet on yourself. Pretend to climb into special space suits that will keep you warm on very cold

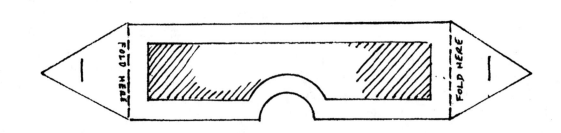

Staple a rubber band length where the "X"s are—make sure each staple has no protruding ends.

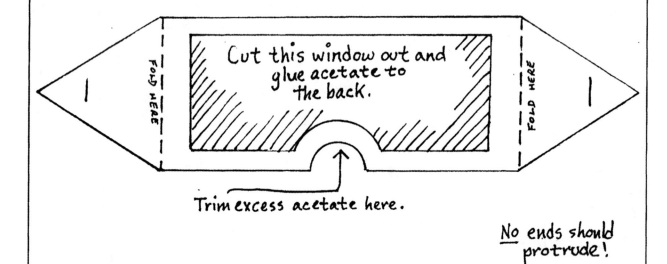

Trim excess acetate here.

No ends should protrude!

Properly stapled staple.

Improperly stapled staple.

After you cut out the first pair of glasses, try them on one of your children. Depending on the age and size of your students, you may want to adjust the size of this pattern, or use the smaller one above.

planets, and cool on very hot planets. Tie the rope to a chair, and explain that when you leave the spaceship, you will all have to hold on to the rope because there is no gravity in outer space, and you do not want anyone to float away. Pretend the calculator is a temperature gauge. Have the children take turns pushing buttons on the calculator to get an indication of the temperature of each planet. Let them use the flashlights on the dark planets. Think of any other "space tools" the children could use. What could they use to steer the spaceship? (Pretend intercoms, pretend cameras for photos to take back to Earth, rocks scattered on the ground as samples—whatever makes it more fun and more real.)

Sit in the spaceship (but not in the front row of chairs—let the children take turns being the leaders of the mission) and ask the children to get ready for takeoff. Together, do a countdown and "blast off." Hold on tight—you are traveling more than 17,000 miles per hour.

Note: What follows is an example of a guided space trip. Feel free to improvise. Be aware that the monologue does not include comments from children. When you actually facilitate this activity, encourage plenty of comments, verbal interactions, and participation as you and the children take the trip together.

Although the following is an unbroken passage about all nine planets, it is a good idea to spread the information over several "space trips" during the unit. You may choose to take one every day, as part of an Attention Getter time. If you visit only a few planets each time, the children will be less likely to confuse the planets because of too much information being introduced at the same time.

TRAVELING IN OUTER SPACE

(After countdown and blastoff) "We are going toward the sun now, and the first planet on our way is Venus. I think I see it out the window. Get ready for landing! Press the cooling button on your space suits—the surface of Venus is almost 900 degrees. Boy, that's hot! The air of Venus is very poisonous, too, so make sure you have your helmet and life-support system on. (Climb out of the spaceship, hold on to the rope, and pretend there is no gravity.) This is the planet that scientists used to call "Earth's Twin" because it is almost as large as Earth. Gosh—look at this planet! No oceans, no lakes, or rivers—no water at all. Would you like living on Venus? Did you know that one single day on Venus lasts the same amount of time as 243 of our Earth days? Wow!"

(Back on the spaceship) "Here comes Mercury! Can you see it down there? Hold on for landing. Mercury is the last planet on our way to the sun. Remember our rhyme: "Mercury's closest to the sun, Pluto is the farthest one." (Climb out onto the surface of Mercury.) Watch out for all these round holes—don't fall in one! There are billions of these craters on Mercury. The biggest one here, Caloris Basin, is 800 miles wide. This planet is nothing but a rocky desert. We're on the dark side of it, away from the sun. This side of Mercury can reach 360 degrees below zero. No place on Earth, not even the North Pole, gets even half that cold. Mission Leader Nicki, what does our temperature gauge say that Mercury's temperature is right now?" (Encourage child to punch buttons on calculator.)

(Back on the spaceship) "Oh my gosh, we're coming up to the sun now. I'm sure glad we have a super spaceship, because if you were on a spaceship that took two days to get to the moon, it would still take you a year to get to the sun—that's how far away it is. But our spaceship traveled the whole 93 million miles in just a few minutes. Put on your space glasses—you don't want to damage your eyes. The sun is the only star in our solar system. Even though we see other stars in the sky at night, none of them are in our solar system—they're all much, much farther away. Oh, the ship is getting closer. Wow! It's such a bubbling, boiling, raging mass of hot, glowing gases. I'm so glad I have these glasses—the light is so brilliant and fierce. I don't think we can get any closer to the sun without burning up—the middle of the sun is 25 million degrees Fahrenheit. Mission Leader Roger, what does the temperature on our temperature gauge show? Oh my gosh!! Look out!! Duck!! It is a huge tongue of gas shooting out—watch out! All stations alert! Whew—that was a close call. That's called a *prominence*, when fire shoots out of the sun like that. Some can shoot out 367,000 miles from the sun. That's the same as if you traveled around the center of Earth fifteen times. That was too close for comfort—we better get out of here. Mission leader, what should we do now?"

(Back on the spaceship) "Now we're traveling away from the Sun, in the other direction, back past Mercury, back past Venus, past Earth (wave out the window). Hi, everyone on Earth! We're exploring the planets in the solar system. We won't be back for a while. There's our moon. People from Earth have been on the moon. Now we're coming to Mars. (Out on the planet.) Mars is only half the size of Earth. See all this rust, or iron oxide, on the ground? (Kick the floor.) That's the reason Mars looks red from far away. There used to be rivers on Mars, but aren't anymore. Does anyone see any dried-up river beds? Oh, I think I see Mons Olympus. See that gigantic volcano over there? It's three times higher than Earth's highest mountain. Oh my gosh, everyone, hold on tight to the rope! (Raise your voice as if you're shouting over the wind.) The wind on Mars can blow with terrific speeds and cause huge dust storms—I think that's what we're in now. Mission leader, should we head back to the spaceship or keep going? "

(Back on the spaceship) "Now we're headed toward the giant planets. Mercury, Venus, Earth, and Mars are very small compared to the other planets. I think I see Jupiter up ahead. It's the fifth planet, but it's not really a planet, because it has no ground to stand on. It's a giant ball of gas. Look at all those swirling storms. Ooh, look at that flash of lightning! See that big red spot? Scientists think that that red spot is a huge, mammoth storm that's so big, it could swallow up three Earths. Jupiter is 1,300 times bigger than Earth, and it takes Jupiter twelve of our years to travel once around the sun. Scientists sent a robot called Voyager to come and take pictures of Jupiter."

"Oh, now we're coming up to Saturn. I can tell it's Saturn because of the rings around it—the rings are made of thousands and thousands of chunks of rock and ice, all floating around it. Does any astronaut want to venture out of the safety of our spaceship to get a rock sample? Saturn is another giant ball of gas. The outside of the gases can get horribly cold, and the center of Saturn's gas ball is horribly hot. Saturn is 800 times bigger than Earth."

"Saturn was the last planet that Earth robot ships have taken pictures of. Now we are traveling where no astronaut or robot has ever traveled before. I'm kind of scared—how about you? Oh—this greenish blue planet is Uranus. It's another gas giant and it is 300 times bigger than Earth. (Take out the skewered orange. Tilt it a little, but with the skewer running vertically, to simulate Earth's axis and the way it revolves. Say: "This is how Earth turns. This part (point to the skewer) is something I put in the orange to show you what an *axis* is—it means the line we would see if we drew one from the North Pole, the top of planet Earth, to the South Pole, the bottom of planet Earth. But this is how Uranus turns. (Turn the orange so that the skewer is horizontal and roll the orange in this position.) It takes Uranus eighty-four Earth years to move around the sun."

"Now we're coming to Pluto. Sometimes Pluto is farthest away from the sun of all the planets, and sometimes it's closer than Neptune. It must be closer than Neptune right now, because we've come to it first. That's because it moves around the sun in a very odd way—sometimes it swings way out into space, away from the sun, and sometimes it moves much closer to the sun. Wow, look at all that ice and snow—I guess Pluto is just one big frozen snowball in outer space. Mission Leader, should we explore it?"

"Now we're coming to the last gas giant in our solar system, Neptune. I know this is Neptune because of the pale blue color. Neptune is 300 times bigger than Earth. What observations do you have of Neptune, astronauts?"

And head back to Earth.

What Would Happen to the Planet Earth without Gravity?

Science/Sensory

WHY we are doing this project: to enable children to experiment with the force of gravity; to help them understand that the gravitational pull of the sun (and the moon) affects the position of Earth in the universe.

WHAT we will need:

    Pillowcases
    Large, heavy balls
    Activity sign (provided below, photocopy
      and enlarge for your use)
    Large tub

HOW we will do it: Put one ball in each pillowcase, and put these "gravity balls" in a tub, near a large open space. (It may be outside). Post the activity sign nearby.

During an Attention Getter time, point to the space poster and remind the children that all the planets move around the sun, and that gravity is the force which makes this happen.

Ask the children what they think would happen to our planet Earth if there was no gravity and if Earth did not travel around the sun. Show the children the gravity balls and the activity sign. Together, read/interpret what the sign says. Ask the children to predict what will happen when they follow the sign's suggestions. Ask the children what will happen if they let go of a ball when another child is standing nearby. Discuss how this can be prevented. (For example, "Make sure there is wide, open space around you before you swing your ball, and

Stand in an open space.
Swing the gravity ball around and around.
Then let the gravity ball GO!!
What happens?
What does it feel like?

that no one else is near you.") Invite them to conduct the experiment by swinging the gravity balls around and around, and then letting them go, as the sign suggests. What happens?

Afterward, point to the sun on one of your posters of the universe. Say: "The sun has gravity, and pulls on Earth. This keeps Earth circling around the sun. This is why all the other planets circle around the sun, too, and why they also have gravity. (Point to the moon.) Even the moon has gravity and pulls on our planet. The pull of the moon makes the water in the sea rise and fall." Encourage observations and descriptions of the children's experience with the gravity ball.

Other ideas you may wish to introduce: Without gravity, planets would have no pattern or predictability to their movements. Planets might collide, or move far away, out of reach of the sun and its light and warmth. Talk about this together. What would planet Earth be like to live on, if this happened? (Make sure the children know that you are just imagining this scenario—it is not likely to ever really happen.)

Rocket Launcher
Science

WHY we are doing this project: to help children understand that there are many different gases; to enable children to make a gas (carbon dioxide) and to see visible evidence of it; to enable children to launch a "rocket ship" which they can pretend is soaring into outer space.

The facts of the matter: The combination of baking soda, which is a base, and vinegar, which is an acid, makes carbon dioxide gas.

CAUTION: Depending on the size of your bottle and the proportions of baking soda and vinegar used, the corks can pop out of the bottles with great force. Depending on the ages of your children, make this project safe by doing either (or both) of the following: Experiment first with your bottles and safe amounts of baking soda and vinegar, which push the corks up with less force, and then make measuring signs suggesting how much of

each material the children should use; or, during an Attention Getter time, discuss with the children how important it is to stand three or four steps back as soon as they put the corks in the bottles.

WHAT we will need:
    Small, clear plastic soda bottles
    Funnels
    Tubs or sensory table
    Two ½ cup measuring cups
    Baking soda (two boxes)
    Small pitchers of vinegar
    Corks
    Paper towels (cut in 12.5 cm [5"] squares)
    Thumbtacks
    Streamers (strips of crepe or tissue paper
        cut in lengths that are 10 cm x 5 mm
        [4" x ¼"])
    Spoons
    Small bowls (for the baking soda)
    Tubs of water (for rinsing bottles)
    Newspaper
    Book: Baird, Anne, *The U.S. Space Camp
        Book of Rockets*
Demonstration:
    Cup
    Baking soda
    Vinegar
    Tray

HOW we will do it: To prepare, attach the streamers to the corks using thumbtacks. Spread several layers of newspaper on your activity table and set out all materials so that the children will have access to them. If you are working with three- and young four-year-olds who are still mastering pouring, you may prefer to do this project in tubs or a sensory table to minimize your clean-up.

During an Attention Getter time, ask the children: "How do astronauts travel to outer space? What takes them there?" Encourage comments. Together, look at and discuss the photographs in *The U.S. Space Camp Book of Rockets*. Encourage the children to describe what they see.

After you have looked at pictures of space rockets, tell the children that today you are going to launch a space rocket yourselves, and that in order to launch it, you are going to use

94

the power of a gas. Ask: "Do you remember what a gas is?" (If necessary: "A gas is not hard, like this table or chair or the floor, and it is not something wet and pourable like milk or water or juice. One example of a gas is the air we breathe.") "We are going to make another gas, now, and it's called carbon dioxide."

Have the children identify the baking soda and vinegar. Ask the children to predict what will happen when they are mixed. Put the tray under the cup, and have children take turns putting small spoonfuls of soda into it. Then have one child pour some vinegar into the cup. What happens?

The paper towel squares in this project are designed to delay the reaction of the baking soda to the vinegar for the time it takes the children to put the corks in the bottle tops. However, this can be problematic for young children who find it difficult to roll something up, like a pile of baking soda in paper. If you think this will be the case for your children, there are several options: make the baking soda packages yourself by rolling ½ cup soda up in a paper towel square and twisting the ends. Make sure the rolls are narrow enough to fit through the bottle opening. Make several of these and put them out with the other materials.

Another option is to cut toilet paper rolls in half lengthwise and line each one with a paper towel square. The curve of the toilet paper may help some children roll up the baking soda packages. You may also provide sticky tape strips to help them seal the rolls.

Show the children one of the corks with the streamers attached. Say: "This is our space rocket. Which planet is it going to travel to?" Prepare a baking soda package and put it into a bottle. Measure ½ cup vinegar and say: "Do you remember what happened when we mixed the baking soda and vinegar? What do you predict will happen to our rocket (point to the cork with the streamer) if I pour vinegar inside our launcher and put the rocket in the bottle opening?" Invite the children to conduct the experiment for themselves. What happens? If necessary, remind the children that the vinegar and baking soda make a gas called carbon dioxide. Using this information, ask the children to hypothesize about why the rocket shoots up after the vinegar is poured into the launcher.

What Do Other Planets Look Like? Making a Spacescape
Art/Crafts

WHY we are doing this project: to stimulate children to imagine what other planets look like; to develop creative expression; to develop fine motor skills; to create interest in the experiment that follows.

WHAT we will need:
Markers
Crayons
Space magazines
Construction paper (black and brown)
Paper shapes (provided on page 96)
Spools
Small plastic aquarium plants
Nontoxic spray paint (silver, gold, and red or orange)
Small stones
Toilet paper rolls
Pipe cleaners
Shallow boxes and lids
Small jewelry boxes
Aluminum foil
Acetate papers
Foil star stickers
Dead coral (small pieces [available in nature stores])

HOW we will do it: The preceding list is comprised of suggestions. The items you collect will depend on your budget and your time. Feel free to substitute items or additional items as appropriate to your circumstances. There are a number of ways to prepare this activity, depending on how much time you wish to devote to it.

Children can make their spacescapes on box lids without backgrounds, or you can cut the sides of boxes down and fit lids onto them so that the spacescapes have backgrounds. You can glue black paper onto the base and background of each box, or leave the boxes the way they are. Do what works for you.

Use different colors of construction paper to make these three-dimensional shapes.

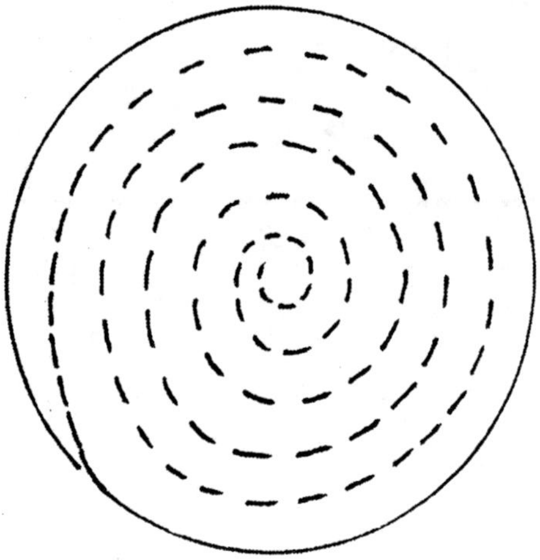

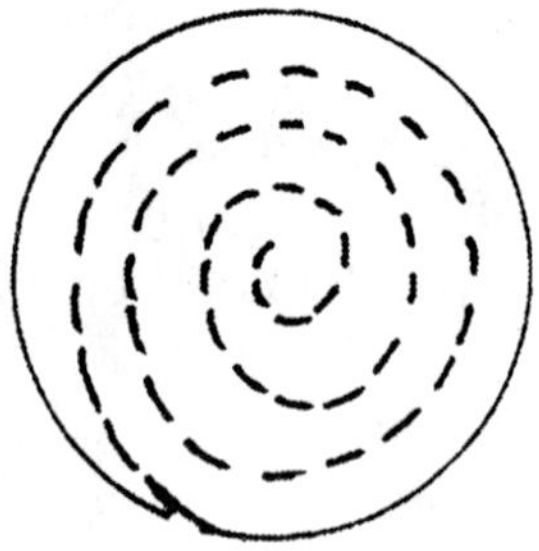

Cut these spirals out, and then pull up the top to make a three-dimensional shape.

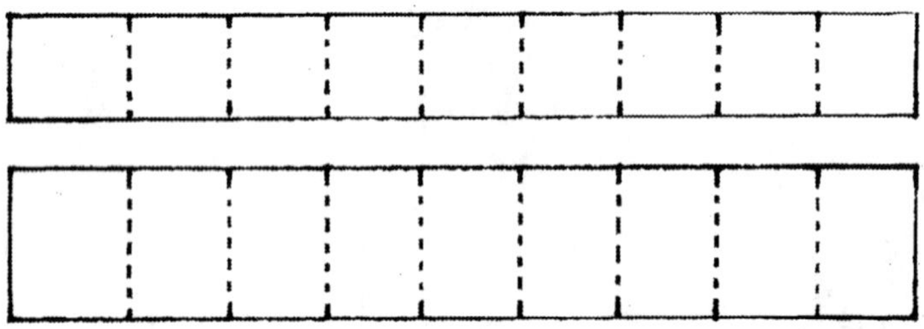

Fold along the broken lines, forward and backward, to make accordian-shaped zig-zags.

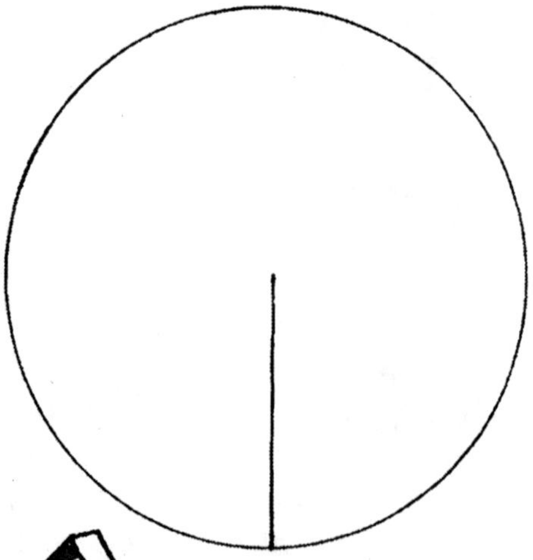

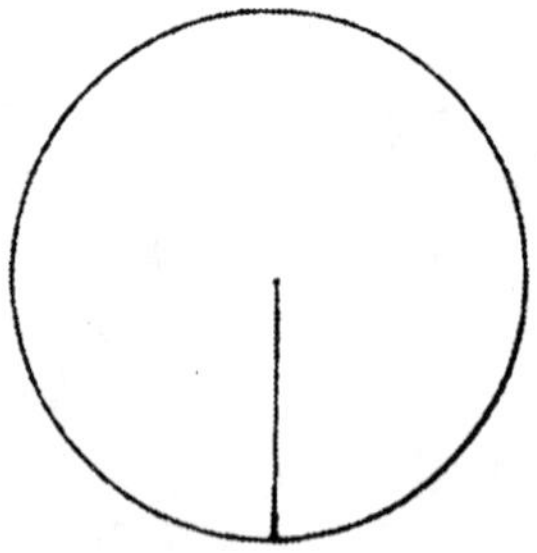

Cut along the broken lines, pull the edges together, and staple to make a disc or satellite dish. Make sure all staples are stapled correctly and that no sharp staple prongs are protruding.

Cut these strips out and fold on the broken lines. Glue the small panel to the outside of the other end to make three-dimensional triangles.

GLUE TO OTHER END

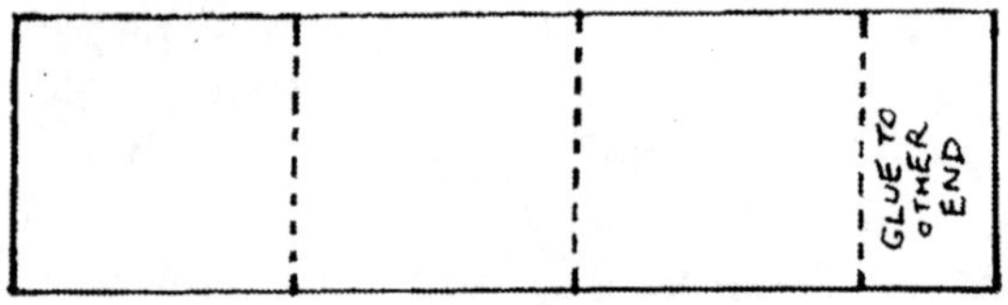

GLUE TO OTHER END

96

To prepare, use the nontoxic gold and silver spray paint to spray some spools, plastic plants, stones, coral, or whatever other objects you want to make "spacey." Do this several days in advance so that the objects have time to dry. Another very "spacey" effect is to spray streaks of red and/or orange paint on black paper. If you would like the spacescapes to include backgrounds, trim the edges of the boxes down to about 8 cm or 9 cm (3") and fit the lids onto the long side of the boxes so that the lids extend upward at right angles from the boxes. Glue the lids to the outside of the boxes. The sprayed black paper can then be glued onto the lids to form backgrounds. (You may have to cut the corners of the lids to fit them onto the box bases.) You can make rudimentary spaceships by decorating toilet paper rolls, bending pipe cleaners, and gluing them underneath the rolls to make "legs" or landing apparatus.

Make a sample spacescape ahead of time. When you are ready for the children to begin the activity, put all materials out on the activity table. Some space magazines have drawings and articles about imagined societies or landscapes on planets in far-off galaxies. *Astronomy* and *Sky and Telescope* are two possibilities. Try to find magazines like these, and during an Attention Getter time, show the pictures to the children and encourage comments. Afterward, take out your spacescape and show it to the children. Talk about your imaginary planet. Ask the children what the name of the planet might be, and if you like, talk together about the imaginary galaxy it is in. Ask the children if they see anything in the room that will help them make their own spacescapes. Be sure to mention that the sample you have is just one way of making a spacescape, and that theirs will look different because they have their own way of making things.

After the spacescapes have dried, display them against a large space poster if you like.

The Crystal Planet
Science/Small Group Activity

WHY we are doing this project: to extend the idea of an imagined spacescape and to allow children to observe the chemical reaction which turns charcoal into crystals.

WHAT we will need:
  Charcoal briquettes
  ¼ cup water
  Large plate
  Tablespoon
  ¼ cup salt
  ¼ cup laundry bluing
  1 tablespoon ammonia
  Food coloring
  Medicine dropper
  Small bowl
  Space poster
  Masking tape

HOW we will do it: During an Attention Getter time, after doing the previous activity, tell the children that you are going to try to grow a crystal planet. Ask: "Do you think there are any crystal planets in another galaxy, somewhere far away in outer space?" Ask the children what it would be like to live in such a place.

Have the children help you put the charcoal briquettes in a bowl. Mix the water, laundry bluing, salt, and ammonia, and pour it over the charcoal. The charcoal pieces should be only partially immersed in the solution. Have the children take turns using a medicine dropper to place drops of food coloring at different places on the charcoal. Put the bowl in a place where it will not be disturbed. Use the space poster as a backdrop, behind the bowl.

The temperature and humidity in your room will influence the speed with which crystals begin to form. Sometimes it only takes a few hours for the first crystals to begin growing. Make a masking tape line on the floor from which children can clearly see the crystal planet, but cannot touch it. The crystals are extremely delicate and will collapse at the slightest touch. Discuss this as a group. Why did the crystal planet start to grow? As appropriate, explain that the charcoal briquettes are full of tiny spaces, and that the water seeped into these spaces. Soon the water in the bowl begins to dry up because the air makes the water into vapor, and the vapor mixes with the air. (Evaporation.) The salt in the water is left behind and forms tiny crystals on anything solid. When water evaporates from the ends of the first small crystals, new crystals grow on the old, which is why the crystals "grow."

Brainstorm with the children together to come up with a name for your crystal planet. Ask: What galaxy is it in? How do you get there? What kinds of creatures might live on a crystal planet?

The Sun and the Planets
Music/Movement/Gross Motor/Small Group Activity

WHY we are doing this activity: to help children "feel" music; to help children understand that the planets revolve around the sun; to develop listening skills; to develop imagination.

WHAT we will need:
    "Spacey" music (see suggestions below)
    Books with realistic pictures or
       photographs of planets
    Tape recorder
    Large, open space
    Flashlights
    Crepe paper (red, green, and blue)
    Tape
    Coffee can with sand

HOW we will do it: In this activity, you are going to create a space-like environment in which children can "revolve" around the sun. To prepare, use the tape and crepe paper to cover the flashlights, so that when they are switched on, their light will be different colors. Arrange them in different places around the room. In the coffee can, place one flashlight which has no crepe paper covering and put this in the middle of your room to be your "sun."

There are some wonderful "space" music tapes on the market right now (sometimes called "New Age" music). One of my favorites is "Astral Journey" by David and Steve Gordon, produced by Sequoia Records, Box 280, Topanga, CA 90290. This is an excellent tape and ideal for this activity. "Celestial Suite" is also produced by these same artists.

Designate certain toys and a certain area for use by children who do not wish to participate in this activity.

When you are ready to begin, take your books or magazines and look at the photographs and pictures of planets in outer space. Ask: "What do you think it sounds like in space? Do you think the planets revolve around the sun quickly or slowly? Do you think there's any life in outer space?" Tell the children that you are going to pretend to be the planets. Turn all overhead lights off, and all flashlights on. Ask your students to guess what the flashlight in the middle of the room is supposed to be. Turn on your space music and revolve around the "sun" as lazily and slowly as planets would in outer space. After a minute or two, ask the children which planets they are. Next, ask the children to imagine that they are astronauts who are exploring outer space. They are the only ones on their space mission who are brave enough to climb out of the spaceship to do this. Cords connect them to the spaceship, but they are slowly floating in outer space. After a few minutes, ask them to imagine that the gravity of a strange new planet is starting to pull them toward it, closer and closer. They land on the surface. It is a planet in another galaxy, and it is the strangest, oddest place they have ever seen. Invite them to explore the strange planet. If you like, take this activity in other directions of your choice. One nice option is to invite the children to lie down and listen and relax to the music for a few minutes.

What Is a Meteor?

Science/Sensory

WHY we are doing this experiment: to help children observe how craters are formed by meteors; to help them understand what meteors are; to provide a sensory experience.

WHAT we will need:
  Any book on outer space that
    describes meteors
  Dirt
  Small pitchers of water
  Spoons (large and small)
  Medicine droppers
  Popsicle sticks
  Pebbles
  Stones (large and small)
  Smocks
  Tubs
  Soapy water and paper towels
    (for clean-up)

HOW we will do it: There is no doubt that this activity can get messy, but children do love it, and it demonstrates in a very fun, hands-on way the impact meteors have on Earth's surface.

To prepare, set out the dirt, stones, tubs, smocks, and pitchers of water in an outside space which you do not mind becoming messy.

During an Attention Getter time, take out your book which features meteors, point to the pictures, and explain: "Sometimes chunks of rock and metal shoot across outer space. When they are out in space, they're called *meteoroids*, but when our planet Earth's gravity pulls them into our sky, they're called *meteors*. They travel so fast, and are sometimes so big and heavy, that when they hit the Earth they make *craters*. One of the biggest craters is in Arizona. It is called the Barringer Crater, and it is almost one mile wide. When a meteor hit that spot, 20,000 years ago, it blasted 400 million tons of rock into the air." Reassure your children that meteors do not crash onto Earth very often, especially not big ones, and that this is not something they have to worry about.

Tell the children that you are going to do an experiment which will show you how meteors make craters. Show them the materials you set out. Ask them to predict what will happen if they make a wet muddy mixture in the tubs with the dirt and water, and then throw mud balls or stones into it. Before you invite them to conduct this experiment, ask: "Is it okay to throw a stone or mud ball at another person? What could happen if you did? Where is the *only* place that it is okay to throw a stone or mud ball?" As the children conduct the experiment, encourage them to observe and verbalize what happens. (If the mud is too thick, they may have to pick the object out before observing the crater, which is why small amounts of water, and small dispensers like medicine droppers and spoons, are recommended.)

Describe meteor showers to the children. (A meteor shower is hundreds of meteors falling in one hour, all coming from the same direction.) Again, reassure your students that this is a rare occurrence, and not something they should worry about. Invite the children to discover the impact of a shower of small pebbles on the mud.

Three, Two, One, Blast Off!

Dramatic Play/Language/Art

WHY we are doing this activity: to stimulate expression of ideas, imaginative play, and child-to-child interaction; to facilitate creative expression.

WHAT we will need:
  Refrigerator box
  Posters of galaxies or the Milky Way
  Steering wheel
  Computer keyboard
  Flashlights
  Chairs
  Duct tape
  Markers

Paper
Glue sticks
Children's scissors
Space theme stickers
Book describing rockets and spaceships
Preparation:
Exacto knife
Markers

HOW we will do it: Obtain an empty refrigerator box from a store that sells large appliances. Using the Exacto knife, cut a door in one side of the box and windows with flaps which can be open or shut in two other sides. Since the box is doubling as a spaceship, you may want to cut circular windows like portholes. (More difficult to cut out, granted, but anything for realism!)

The items listed are just suggestions; if you can think of additional or substitute items, use them to create a rocket ship that will feel as real to the children as possible. In our spaceship, we used duct tape to secure a flashlight onto the ceiling, and put a poster of the galaxy in front of the steering wheel, to simulate the view into outer space. Whatever you can find to simulate a control panel—calculators, computer keyboards, or Fisher Price toys with knobs and levers—will greatly enhance the activity.

On the first day of this activity, ask the children for suggestions about what the rocket ship should be called. Write the name on the outside of your rocket ship. Display the book on rockets and spaceships, and make a point of observing the letters, words, and symbols on the outside of the spacecrafts. Discuss these with the children and encourage their comments. Show them the paper, scissors, markers, glue sticks, and stickers. Invite them to use these materials to create a spaceship exterior.

Put the space helmets, life-support systems, and space glasses out near the spaceship, then stand back and enjoy!

Night Sky
Sensory

WHY we are doing this project: to provide children with a sensory experience which utilizes the theme of the unit.

WHAT we will need:
Empty cereal box
Flashlight
Sturdy paper plates (cardboard squares about 21 cm [8"])
Preparation:
Scissors
Exacto knife
Skewer

HOW we will do it: In this project, you will be punching small holes into the bottom of a cereal box, and securing a flashlight inside. When the flashlight is turned on in a dark space, and the light shines through the holes, the effect simulates a starry sky.

To begin, use the scissors, skewer, and Exacto knife to punch holes into the bottom of the box. Star shapes produce a wonderful effect, but are quite difficult to cut. Next, fold the paper plate or cardboard in half, and make a small horizontal cut. Then fold it the other way and make another horizontal cut so that your cuts form a cross (+) in the center. Push the flashlight through it. Bend the edges of the plate or cardboard so that you can wedge it into the box. (Because the sizes of flashlights and cereal boxes vary, you may have to experiment with the best way to secure your flashlight in your box.)

Make several of these and create a dark space; you may be able to use your refrigerator box for this purpose. Invite the children to use them.

Moonshine

Science/Language

Note: This project requires parent participation.

WHY we are doing this project: To encourage children to observe the moon and to develop language skills.

    Tip: It is better to do this activity during winter if possible because in summer, the moon may not be visible until after bedtime. If the moon is visible during school hours, take the children outside to observe it.

WHAT we will need:
    Book: Asch, Frank, *Moongame*
    Paper
    Pens
    Markers
    Butcher paper

HOW we will do it: Put your butcher paper up on the wall and have a marker handy. During an Attention Getter time, read *Moongame*. Afterward, ask the children to tell you about a time when they saw the moon. Tell them that you have some important homework for them to do. You would like them to observe the moon. Ask: "When do you think would be the best time for getting a good look at the moon?"

    There are several variations on this activity. You may ask your students to look at the moon for one night only, and then to describe their observations the next day during group discussion. You may ask parents to take story dictation from the children describing their observations. You may ask the children to observe the moon over a week or so and to draw pictures of what they see so that they can observe how the moon appears to change shape. If the children record their observations with words and drawings, have a group show-and-tell of their work. If you work with three- or young four-year-olds, have only three or four children show-and-tell their work at one time.

    Whichever option you choose, write a letter to parents explaining what you are doing and asking for help with this special homework assignment. Take dictation and write the children's words for the letter on the butcher paper.

If you work with kindergartners, they may want to write their own individual letters. If necessary, send your own letter with the children's. Include a suggestion that children notice whether there is any light in their yard or house that comes from the moon.

Observatory

Science/Dramatic Play/Language

WHY we are doing this activity: to enable children to understand what astronomers do and the tools that they use; to expand vocabulary; to facilitate imaginative play and child-to-child interactions.

WHAT we will need:
    Cardboard poster tubes
    Chairs
    String
    Packaging tape
    Graph paper
    Pens
    Markers
    Table
    Chairs
    Measuring devices (found in geometry
        sets; e.g., ruler, protractor)
    Computer monitor and keyboard
        (if possible)
    Play bed, table, chairs, dishes—
        astronomers' living quarters
Optional:
    Black construction paper
    Photographs of planets from astronomy
        magazines
    Star-shaped stickers

HOW we will do it: In this activity, you are going to set up a dramatic play observatory. Make telescopes out of the cardboard poster tubes. Remove the plastic ends so children can look through them. If you like, make them look a little more real by wrapping them in aluminum foil or construction paper. Attach the tubes to the backs of chairs, at angles, using string and/or tape, or hang the "telescopes" from the ceiling with string.

Either position your telescopes near windows so that when the children look through them, they see the sky, or aim them toward a sky picture pinned up on the wall. You may wish, as a group, to make your own picture of the sky or solar system to place in front of the telescope. To simulate a night sky, put photographs of planets, star stickers, or whatever else you like onto black construction paper.

Put the table and chairs near the telescopes. Put the graph paper, rulers, pens, and markers on the table. Set up the computer and monitor nearby. Depending on the computer and programs you have available, you may decide to have the children enter their data with the computer on, or you may leave it switched off and simply let the children pretend while typing on the keyboard.

In another part of the "observatory," use the play furniture and dishes to set up the astronomers' living quarters.

Look through your astronomy magazines to find a drawing or photograph of a telescope. During an Attention Getter time, show the telescope to the children and ask them what they think it might be for. Let the children know that people who look into the sky to learn about planets, stars, meteors, and comets are called *astronomers*. Astronomers study *astronomy*. Say these words several times and pat your head, blink your eyes, or tap your feet on the floor in time to the syllables. Show the children the dramatic play area and let them know it is the "observatory." Explain that an *observatory* is where astronomers study the sky. Show the children the play furniture in your observatory and ask them why they think astronomers might need a place to sleep and eat while they work. Discuss the work that astronomers do; for example, predict what will happen in the sky, *chart* the paths of planets, or take photographs or draw pictures of the planets to learn about their size and position. Show the children the materials on the table for these purposes and invite them to explore. Use the new vocabulary words as much as you can during the day: *astronomer, astronomy, observatory, chart*.

Space Books
Science/Language/Art

WHY we are doing this project: to develop all language arts skills: reading, writing, speaking, and listening; to expand vocabulary; to teach children about other things connected with space technology.

WHAT we will need:
    Space magazines (e.g., *Astronomy,*
      Sky and Telescope)
    Paper
    Glue sticks
    Children's scissors
    Markers
    Crayons
Preparation:
    Stapler
    Scissors
Optional:
    Puppet

HOW we will do it: To prepare, take two or three sheets of paper stacked on top of each other, and fold them in half. Staple twice along the fold to make a blank book. Make one of these for each child. If you work with three- and young four-year-olds, cut out some magazine pictures (e.g., asteroids, comets, meteoroids, space robots, satellites, rocket ships, astronauts, telescopes, constellations, and planets). Older children can do this themselves. Put out all materials, and include blank sheets of unstapled paper.

Make a sample space book ahead of time by gluing pictures in your book and writing about them. Examples: (Realistic magazine drawing of red planet.) "This planet looks spooky. It is red, like Mars. It looks cold and lonely to me." (Drawing of spaceship.) "This is the kind of spaceship I'd like to travel in. In the refrigerator, I drew my favorite food: grape juice and potato chips. I would eat them in outer space."

During an Attention Getter time, read your book to the children. Ask them if they see anything in the room that would help them make their own space books. As the children use the materials, reinforce the names of the space-related objects in the magazine pictures through conversation, or use your puppet to do this. Take story dictation, write down words to be copied, spell words, or support invented spelling, depending on the needs of your children. As and when appropriate, explain the new things in the pictures.

At the end of the day or session, or even the next day, have a show-and-tell of space books.

Space Math
Math/Multicultural/Anti-Bias

WHY we are doing this project: to practice rational counting and subtraction; to facilitate sorting into sets; to create multicultural, anti-bias awareness; to develop self-esteem and a sense of autonomy through use of a one-person work station.

WHAT we will need:
Large cardboard panel (from large appliance box)
Black paint (or black contact paper)
Clear contact paper
Cardboard
Space shapes (patterns provided on page 104; photocopy and enlarge for your use)
Double-sided tape
Writing sheets (provided on page 105; photocopy and enlarge for your use)

Blank paper
Pens
Crayons
"One person may be here" sign (provided on page 165; photocopy and enlarge for your use)
Preparation:
Scissors

HOW we will do it: To prepare an "outer space" board, cut off the ragged ends of the cardboard panel. Paint it with black paint, or cover with black contact paper, then cover again with clear contact paper. Prop the board against a wall.

Make several copies of the shapes provided, and color them. If necessary, ask friends or parents to help you with this. Using double-sided tape, stick all the shapes to cardboard, cover on both sides with clear contact paper, and cut them out. Put a few strips of double-sided tape on the back of each shape, and stick them to the black "outer space" board. Put copies of the writing sheet, blank paper, crayons, and pens near the board. Post the "One person may be here" sign on the wall above it.

During an Attention Getter time, show the children the board. Together, as a group, begin to sort the different shapes by grouping them in different places on the board, but do not finish this process. (Leave that for the children to do.) Together, read/interpret a writing sheet, and discuss the "One person may be here" sign. Show the children the blank paper, and let them know that if they like, they can use this to record the number of shapes they count. Encourage exploration!

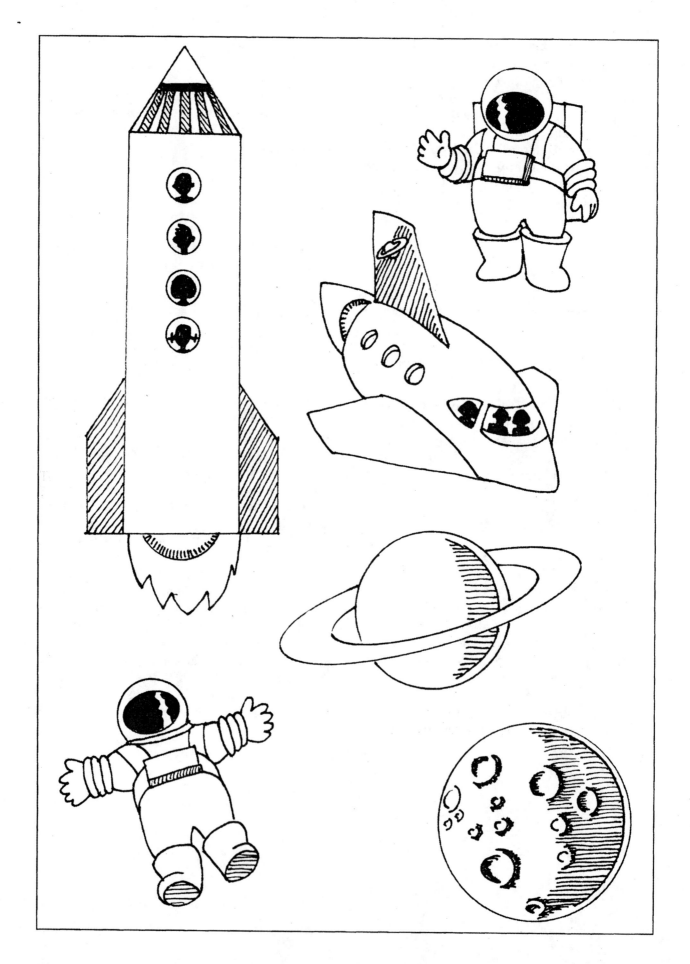

104

How Many?

How many rocket ships
do you count? _____

How many ringed planets
do you count? _____

How many astronauts
do you count? _____

2:
Take two astronauts away.

How many do you count now?

_____

Literature

Symbol Key: *Multicultural
              +Minimal diversity
              No symbol: no diversity or no people

Asch, F. (1984). *Moongame*. New York: Simon & Schuster.

Baird, A. (1994). *The U.S. Space Camp Book of Rockets*. New York: Morrow Jr. Books.

Cole, J. (1990). *The magic school bus lost in the solar system*. New York: Scholastic.*

Cother, J. (1990). *Sky above, earth below*. New York: Harper & Row.

Fernandes, K. (1991). *Zebo and the dirty planet*. Buffalo, NY: Annick Press.*

Fradin, D. (1984). *Skylab*. Chicago: Children's Press.

Hansen, R., & Bell, R. A. (1985). *My first book of space*. New York: Simon & Schuster. (Excellent photographs.)

Ride, S., & O'Shaughnessy, T. (1994). *The third planet*. New York: Crown. (Good reference picture book.)

Extenders

Language: After the children make their spacescapes, encourage them to make up names for their planets. Take story dictation or have the children write about their planets. Ask open-ended questions to get the ball rolling: Who or what lives on your planet? What kind of air or gas surrounds your planet? What grows on your planet? What kind of lightness, darkness, warmth, and cold are on your planet?

Language/Cognitive: To add language development to the Space Math activity, print the name of each object on cardboard rectangles. Place a reduced picture of the object beside the name. Cover the name cards with contact paper and put double-sided tape on the back. Invite the children to match each word with the object it names.

Art: Borrow a book from the library about rocket ships and, as a group, look through it together. Tape several pieces of butcher paper together and cut one long rocket shape. Encourage the children to paint and/or draw on it. Cut slashes in several pieces of colored acetate and glue them under the tail of your rocket ship to simulate fire explosions from blastoff.

Another variation is to decorate two identical rocket ship shapes, staple or glue them together, stuff the inside with newspaper, and hang it from your ceiling. If you do this, staple the middle edges together first, before you fill that section with stuffing. Then staple the remainder, moving toward the ends as you stuff.

FLUID FORMS

In this unit, children will explore the comparative densities of different fluids. Liquids you can use for this unit include: honey, vegetable oil, dish detergent, water, syrup, hair gel, yogurt, milk, or cornstarch liquid.

Test tubes are a lot of fun to use with young children because they really evoke the feeling of being a genuine scientist. If you can find plastic test tubes in catalogs for older grades, they are a good investment. Do not forget to order a rack to hold them in, as well.

Attention Getter: Gather the children together and ask them to turn their backs to you. When they cannot see you, take out a tub and a big pitcher of water. Tell the children that they are going to hear a noise and that they have to guess what it is. Pour the water so that it flows very quickly and then pour it so that it trickles. When the children guess that they are hearing water being poured, ask them to see if they can hear another fluid being poured. Tip a bottle of vegetable oil into the tub. Ask the children how it sounded: the same as the water? Why not? Put a small container of each with pouring cups inside a sensory tub for later exploration. Tell the children that there are two other names for things that are wet and can be poured: fluid and liquid. You may want to introduce these words on different days. Tap your fingers or feet, and blink in time to the syllables as you say the words.

Pouring Fast and Pouring Slowly
Science/Sensory

WHY we are doing this experiment: to help children understand the connection between the thickness of a liquid (degree of density) and speed of motion; to provide a sensory experience; to develop fine motor coordination (specifically pouring skills); to develop all components of language arts: speaking, listening, reading, and writing.

The facts of the matter: Everything is made of *molecules*. A dense liquid, like honey or syrup, has more molecules packed together than a liquid like milk.

WHAT we will need:
 Sensory table or large tubs
 Newspaper
 Small pitchers
 Containers
 Cups
 Fluids (thin and thick: e.g., syrup, liquid
  soap, water, cornstarch liquid, honey,
  molasses, and milk)
 Butcher paper (two pieces)
 Marker

HOW we will do it: To prepare, spread newspaper underneath the work area, set out all the materials, and pin up the pieces of butcher paper close to the place where the children will be conducting the experiment. On one chart,

print: "Prediction Chart." On the other, print: "Results of Our Science Experiment." Divide both papers into two columns, each with one heading that says: "Fluids that pour quickly— runny fluids" and another which says: "Fluids that pour slowly—dense fluids." Read or interpret these headings together and discuss the meanings of *dense* and *runny*. Show the children each fluid and ask them to predict which will flow quickly and which will flow slowly. On the prediction chart, print (or have the children write) their predictions. Then ask how the children can find out for sure how the liquids flow. Encourage experimentation, and invite the children to record their findings on the results chart. After the experiment, read the charts together and compare them. Talk about how the dense or thick liquids pour slowly and how thin or runny liquids pour quickly. To reinforce the meaning of *dense*, discuss which is more dense: the honey or the water, the syrup or the milk.

The Oil Experiment
Science

WHY we are doing this experiment: to enable children to observe how oils and water-based fluids separate.

WHAT we will need:
 Transparent plastic containers with lids
 Vegetable oil
 Sesame oil
 Red vinegar
 Soy sauce
 Small pitchers
 Funnels
 Large sensory table
 Small containers
 Spoons

HOW we will do it: Put all other materials in the sensory table and spread several layers of newspaper on the floor underneath. Sesame oil is expensive so use small amounts. It is darker than vegetable oil and is easy for children to distinguish, which is why it is necessary for this experiment. Have on hand sesame oil, vegetable oil, soy sauce, and red wine vinegar (in their original bottles) for your Attention Getter demonstration.

During Attention Getter time, identify each fluid with the group, and then put some of the oils, vinegar, and soy sauce in a container and rest the lid on top of the container. Say to the children: "What do you think would happen if I shook this jar really hard right now, without screwing the lid on the container?" Talk about how important it is to screw the lid on tightly and evenly. Do this, and then pass the container around to the group of children so that each child can shake the container. Set the container on a surface and ask the children to predict what will happen to the fluids if you let the container sit still for a few minutes. Ask the children what they see in the room that would help them to conduct this science experiment for themselves.

Encourage experimentation, and together, discuss the results. Ask: "How do the oils feel? What do you notice about which fluids feel the same? What do you see after the fluids settle in the containers?" Ask the children to hypothesize why the vinegar and soy sauce remain mixed, and why the two oils do the same. Do the oils and water-based fluids all stay mixed up together?

Sinking and Floating in Different Fluids

Science

WHY we are doing this experiment: to enable children to observe that the same object may sink in one kind of fluid and float in another.

WHAT we will need:
> Water
> Syrup
> Vegetable oil
> Small pitchers or plastic cups
> (for pouring fluids)
> Clear plastic containers with lids
> Small plastic objects (e.g., manipulative
> blocks, pen caps)
> Grapes
> Corks
> Carrots cut into grape-size pieces
> Large sensory table

Note: If you work with three- and young four-year-olds and are concerned that the children may put the small objects in their mouths, you may decide to omit this activity or to facilitate it as a group project when you can supervise closely. Another option is to recruit another adult for the session to help supervise. The children do enjoy this experiment.

Some syrups separate from water after being mixed with it, and some syrups do not. Even if your syrup and water mix, you will find that the grape and carrot sink through the oil while the cork and some plastic objects float on its surface. This means that even if your syrup and water do not separate, the children will still be able to observe how the same objects sink or float in different fluids.

HOW we will do it: Spread many layers of newspaper underneath your activity area. Put all materials in the sensory table or in basins.

During an Attention Getter time, pour some oil, syrup, and water into a clear plastic container. Show the children a grape, carrot piece, cork, and plastic object. Ask them to predict what each one will do in each kind of liquid. (If you like, make a prediction chart. Draw a picture of the original container of each fluid at the head of a column, and along the sides, the floating or sinking objects. Write down the children's predictions in appropriate space or invite them to do so.) Ask the students what they see in the room that will help them conduct this experiment. Facilitate experimentation, and together, discuss the children's findings. Ask your students what happens when they screw on the container lids evenly and tightly, and shake the containers hard. Does it change how the objects sink and float?

Air and Density
Science

WHY we are doing this project: to allow children to observe how air moves through dense liquids; to help develop self-esteem and a sense of autonomy through use of a one-person work station.

WHAT we will need:
>   Three plastic containers (empty peanut
>   butter containers are perfect)
>   Liquid soap
>   Water
>   Honey
>   Straw
>   "One person may be here" sign (provid-
>   ed on page 165; photocopy and enlarge
>   for your own use.)
>   Magnifying glass
>   Flashlight
>   Large, strong cardboard box
Preparation:
>   Exacto knife

HOW we will do it: Pour a generous amount of honey in one container, water in the other, and liquid soap in the third. Put the magnifying glass on the table beside them, and pin up the sign on the wall above the table. Cut a three-panel divider out of your cardboard box so that it is freestanding.

Use the Exacto knife to cut a + shape in the middle panel, and push the flashlight through. Use string or duct tape to help you secure the flashlight if necessary. The light should be positioned so that it will be shining into the containers.

During an Attention Getter time, show the children the materials. Tell the children that the density of a liquid affects how quickly or slowly air bubbles move in it. Show them the magnifying glass, and tell them they may use it to observe the speed of the air bubbles in the fluids. Ask them to predict whether air bubbles will be moving quickly, moving slowly, or not moving at all in each fluid. Discuss the meaning of the "One person may be here" sign. Right before the group disperses, blow lots of bubbles into each container. Make sure there is plenty of extra space in the honey container, or the honey will overflow when you inject bubbles in it. Encourage the children to observe the fluids, and as each child observes the experiment, ask questions like: "What do you notice about which direction the bubbles are floating? What do you notice about which fluid has the most bubbles in it? What do you notice about how fast or slow the bubbles are moving?"

Force and Density
Science

WHY we are doing this project: to reinforce the concept of *density*; to give children the opportunity, through their own hands-on experimentation, to discover the connection between the degree of force needed to blow bubbles into a liquid and the degree of density of the liquid.

WHAT we will need:
>   Honey
>   Water
>   Syrup
>   Apple juice
>   Small, clear plastic cups
>   Small pitchers

Straws
Trays
Wastepaper basket
Popsicle sticks

HOW we will do it: Set out the trays on the activity table and, on each one, put three cups. If you can only find large plastic cups at the grocery store, cut them down with scissors. This is so that each child only uses a small amount of each fluid. Place the wastepaper basket near the table. Put a pitcher of each fluid on the table so that all children will have access to them.

During an Attention Getter time, take two cups and a straw. Ask the children if they remember observing the speed of bubbles in different fluids (previous activity). Tell them that they can try to blow bubbles into fluids themselves, and this time, you hope they will notice how hard or easy it is to blow into the different liquids. Sit on the floor so that the children can see into the cups, and pour the first one half full of juice. It is important to pour the liquids while the children watch so that they can observe the difference in density as the liquids flow quickly or slowly. Ask the children to predict whether or not they will be able to blow air through a straw into the juice. Take another cup and pour honey into it. Ask the children to predict whether or not they will be able to blow air through a straw into the honey. (The level of the honey rises quickly as air is injected into it, but it is difficult to create bubbles in it.) Ask the children what they see in the room that will help them to conduct this science experiment.

Hand a straw to each child, and then ask: "Is it okay to suck up the fluids through the straw instead of blowing bubbles into it? Why not?" (There will not be any fluids left for other children to experiment with and germs will be spread.) Children may take a taste anyway, but it is important to let them know that that is not the purpose of the activity. If you like, use the Popsicle sticks to give each child a taste of the honey and syrup before the activity.

In addition, be sure to emphasize to the children that everyone has their own straw, which should be thrown away when they are finished with the experiment. Ask: "What will happen if you leave your straw in the cup and someone else uses it?" (Germs will spread.) "What will happen if you have a cold, and you leave your straw in the cup and someone else uses it?" (That person might get your cold.)

Encourage experimentation and communication of results. Ask the children what they notice about how easy or hard it is to blow bubbles into the different fluids. Ask why they think it is easy to blow bubbles in some fluids and not others. Say: "The honey/syrup is very *dense*." Discuss how a dense liquid has more molecules packed together than a watery liquid like the juice.

What Is a Suspension?
Science/Sensory

WHY we are doing this experiment: to provide children with a hands-on exploration of a fluid which is a suspension (they can learn that a suspension contains very small particles); to provide a sensory experience.

WHAT we will need:
Tomato juice
Strainers
Spoons
Coffee filters
Tubs or sensory table
Clear plastic containers
Demonstration:
Apple juice

HOW we will do it: To prepare, put all materials except the apple juice into the tubs or in the sensory table. During an Attention Getter time, take a clear plastic container of apple juice and a clear plastic container of tomato juice and hold them up, side by side, so that the children can clearly see both. Ask them what the difference is between them. There are several things the children may mention, but if or when someone mentions that the apple juice is clear, or see-through, and that the tomato juice is thick and cannot be seen through, tell your students that the tomato juice is a *suspension*. A suspension has tiny particles in it. Tomato juice is a suspension because it contains a lot of tiny tomato particles.

Ask the children what they see in the room that will help them prove this. Encourage exploration of the materials. What is left behind when tomato juice is forced through a strainer? What can be seen when it is poured through a coffee filter? Before you clean up for the day, invite the children to pour some tomato juice into a clear container, and leave it for a day or two. What happens? Did the particles settle?

The Liquid Book
Science/Language/Art/Fine Motor

WHY we are doing this project: to reinforce the concept of density in relation to fluids; to develop all components of language arts: reading, writing, speaking, and listening; to facilitate creative expression; to develop fine motor skills.

WHAT we will need:
   Food magazines
   Pale-colored construction paper
      (large sheets)
   Dark-colored construction paper
   Index cards
   Scissors
   Glue sticks or glue and brushes
   Markers
   Crayons
   Pens
Preparation:
   Stapler
   Aluminum foil
   Sample shapes (provided on page 113;
      photocopy and enlarge for your use)

HOW we will do it: Begin collecting food magazines well ahead of time; ask friends, family, and other parents to help. To make a blank book, fold three sheets of construction paper over and staple along the fold.

Using the pattern provided, cut the fluid shape out of aluminum foil. Cut the pitcher shape out of dark-colored construction paper. Use a glue stick to glue the aluminum foil fluid shape onto the cover of the book, and then to glue the pitcher shape above it so that it looks like the water is being poured out of the jug. On the top write: "The Fluid Book."

Cut out enough fluid and pitcher shapes for all the children. Leaf through food magazines, and tear out pictures of any kind of fluid: juice, syrup, cream, soup, etc. If you are working with very young children, cut some of these out of the pages and leave some pages intact. Older children can cut their own shapes out of the magazine pages.

A few days ahead of time, make a sample fluid book. Cut out photographs from the magazines of fluids that appeal to you and glue them into the pages. Print words for each photograph. Your book might be as simple as: "This brown liquid is gravy. I like gravy on my mashed potatoes. This is another brown fluid, but this one is syrup. I bet it is sticky to touch." On the cover, under the pouring pitcher, write: "by (your name)."

Set the blank books, magazine pictures, pitcher and fluid shapes, glue, scissors, markers, and crayons on the activity table. Provide aluminum foil sheets and construction paper for children who want to cut their own shapes for their book covers. On the index cards, print a descriptive word on each one. Use words like: runny, sticky, drizzle, gooey, thick, and thin. Put these on the table with blank cards and pens. Children may choose to copy them, color them, or cut them out.

During an Attention Getter time, show the children the cover of your book and read what it says. Then read the rest of the book. Ask the children what they see in the room that would help them make their own fluid books. If you are working with three- and young four-year-olds, take story dictation and/or encourage their scribbling. If you are working with older children, help them in the way that is most appropriate to their needs or requests: spell words, write their words down for them to copy into their books, or support invented spelling. At the end of the day, gather together

112

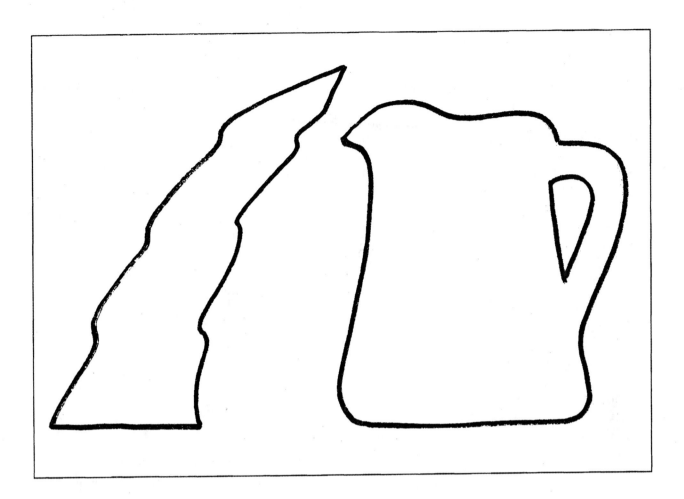

as a group, and have a show-and-tell of fluid books. If there are some children who didn't make fluid books, let them show-and-tell whatever else they prefer.

Weight, Force, and Density
Science

WHY we are doing this project: to help children understand the relationship between the density of a liquid and how that liquid absorbs a heavy, speeding object; to develop fine motor skills.

WHAT we will need:
Clear plastic cups
Trays
Pitchers
Honey
Hair gel
Syrup
Water
Small and large marbles (steel and glass)
Demonstration:
Two large, clear, plastic containers
Peanut butter
Vegetable oil
Can of clear soda (Sprite, Mountain Dew, etc.)
Two medium-sized rocks (must fit inside containers)
Newspaper

HOW we will do it: To prepare, put several plastic cups and marbles on each tray, and nearby, place the pitchers of fluids so that they will be accessible to all children. To control the amount of fluid each child pours, you can cut the cups down so that they are smaller in height.

For the demonstration, make some peanut butter more runny by stirring some vegetable oil into it. It should be runny enough to pour out of a pitcher, but thick enough so that the rock does not sink immediately to the bottom of the container. Spread newspaper underneath your demonstration area. During an Attention Getter time, pass your rocks around the circle and ask the children to feel the weight of them. Put the two clear containers on your tray and, in front of the children, pour the soda into one, and the peanut butter liquid into the other. It is important to pour the liquids as slowly as you can so that the children can observe their relative densities prior to the demonstration. Tell the children that you are going to drop a rock into each fluid, and that you want them to watch and see if the same thing happens to the rocks. (Beforehand, experiment with how high to hold the rock before dropping it. Getting the desired effect depends on the size of the rocks and how much liquid there is in the containers.) After you let the rock fall into the soda, ask: "What happened?" (It should sink immediately to the bottom of the cup.) Drop the other rock into the peanut butter fluid, and ask the children what they see. (It should sink very slowly to the bottom of the cup, or even stop sinking before it reaches the bottom.) What did the children notice about the speed with which the rocks travelled in the fluids? Why did one rock sink quickly and the other slowly? Which fluid is more dense?

Show the children the marbles on the activity table and pass some around the circle so that the children can feel their weight. Bring a pitcher of each substance from the project table to the circle and ask the children what is inside each one. After the group has identified the fluids, ask the children to predict how the marbles will travel when they are dropped into each fluid. Ask the children how they could find out. Invite them to discover more about weight, height, speed, and density with the materials on the activity table. Before they disperse to explore, talk about the fact that the liquids are for dropping objects into, not for drinking or tasting. Together, discuss their findings.

Funnel Experiment
Science/Sensory/Math/Fine Motor

WHY we are doing this project: to help children understand the relationship between the density of a liquid and its ease of movement; to provide a sensory experience; to familiarize children with measurement cups and their relative volume; to practice rational counting; to provide practice with pouring.

WHAT we will need:
  Sensory table
  Large basin or tub
  Honey
  Corn syrup
  Colored water
  Cornstarch liquid (cornstarch mixed with water)
  Small pitchers
  Measuring cups (several complete sets)
  Funnels
  Empty shampoo bottles (or other bottles that are tall with narrow openings)
  Tub of warm, soapy water
  Large bowl
  Writing sheets (provided on page 115; photocopy and enlarge for your use)
  Blank paper
  Markers
  Small table for one-person work station
  "One person may be here" sign (provided on page 165; photocopy and enlarge for your use)
Preparation:
  Permanent marker

HOW we will do it: To prepare, use a permanent marker to write in large numbers and words the measurement of each cup on the cup itself. I have not found measuring cups specifically made for children, and on most sets the measurements are very hard to see. By writing the measurements with a bold, permanent marker using large letters and numbers, this project provides much more of a learning opportunity for children.

Place everything in the sensory table except: one set of measuring cups, one funnel,

114

Science Experiment:
Liquids ⁆⁅ and Funnels

Draw a bottle underneath the funnel that resembles in size and shape the bottle you are actually using. Then photocopy this master sheet.

How many quarter cups

fill up the container?

Science Experiment:
Liquids ⁆⁅ and Funnels

How many third cups

fill up the container?

Science Experiment:
Liquids ⁆⁅ and Funnels

How many half cups

fill up the container?

three small-sized plastic bottles, a large bowl, a pitcher of cornstarch water and a pitcher of corn syrup. Put these things in the basin or tub, and place it on the small table. Place the writing sheets, pens, and blank sheets of paper on the small table also, in order to create a one-person measuring/writing project. If you put the writing materials near the majority of materials, they can become so splattered with liquids that they are impossible to use. Also, children may dribble fluids on the floor if they have to go back and forth between a sensory table and writing materials. The writing materials, and the contents of the tub, are intended as an extension of the exploration of fluids and funnels; a one-person-at-a-time measuring/writing project. (It would be too expensive to supply several children with enough honey and corn syrup to fill several bottles.) The bowl in the tub is for emptying out the contents of the bottles so that the next child can begin the measuring project. It does not really matter whether your students use corn syrup or cornstarch water together or separately. The idea here is that they count and measure and, at the same time, notice the difference in how a thick and thin fluid pass through a funnel.

During an Attention Getter time, show the children the different fluids, funnels, and containers. Also, bring the tub with the one-person measuring/writing project to the circle. Together, read or interpret the writing sheet and show the children where the pens and sheets will be, as well as the blank paper for children who want to create their own format for recording results. Show the children how, if they choose, they can make a mark on the corresponding writing sheet for each cupful they pour. Next, use a set of measuring cups and colored water to demonstrate that it takes four ¼ cups to fill up one whole cup, two ½ cups to fill up one large cup, and so forth. Determine how complete or lengthy this explanation is depending on the age of your children. Younger children may not be interested, and older children need to have something left for them to discover for themselves.

Invite the children to explore the materials and, as appropriate, talk to them about their discoveries as they experiment. Which fluid goes through the funnel fastest? Which one passes through most slowly? Why? Show the

children the basin of soapy water for cleaning their materials when they want to begin a new experiment. Do not be surprised if this, too, becomes a material for exploration.

Density As a Barrier

Science

WHY we are doing this experiment: to provide children with a hands-on opportunity to discover that a dense liquid is "stronger" than a less dense one; to promote self-esteem and a sense of autonomy through use of a one-person work station.

WHAT we will need:
> Tray (or newspaper)
> Tub
> Funnel
> Colored hair gel
> Water
> Food coloring (a different color than the hair gel)
> Clear container
> Pitchers
> Spoon
> Activity sign (provided on page 117; photocopy and enlarge for your use)
> "One person may be here" sign (provided on page 165; photocopy and enlarge for your use)

HOW we will do it: On a tray on the small table, place the tub, funnel, small pitcher of hair gel, small pitcher of colored water, and container. Tape the activity sign and the "One person may be here" sign prominently beside the table. During an Attention Getter time, read or interpret the signs together. Ask the children to predict what they will see when they follow the suggestions of the activity sign. As they discover that a glob of hair gel in the funnel will prevent water from passing through it, ask what they notice about the densities of the two fluids and their relationship to this discovery. There are several ways to do this: either during each child's time at the work table, or during a group gathering at the end of the day.

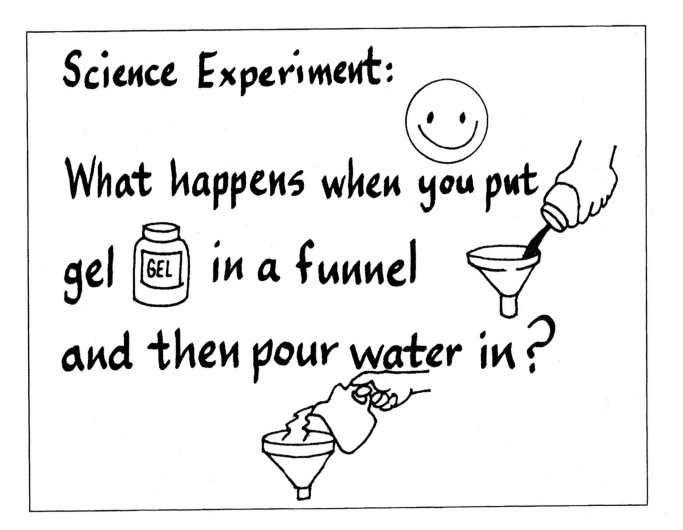

Liquid Drop
Science

Activity sign (provided on page 118;
    photocopy and enlarge for your use)
Puppet

WHY we are doing this project: to enable children to learn, through hands-on experience, that dense liquids do not disperse on impact when dropped from a height and to allow them to compare this to the reaction of watery fluids.

WHAT we will need:
    Small containers
    Medicine droppers
    Hair gel
    Honey
    Cream
    Water
    Newspapers
    Colored construction paper
    Plastic cling wrap

HOW we will do it: The key to this project, in order to avoid a catastrophic mess, is to put very small amounts of each liquid into each container, and refill them as needed. Spread plenty of newspaper out on your activity table. For this experiment, you will need a surface which does not build surface tension in water. For example, even if you drop water from a height onto wax paper, the drops will not splatter. One surface which works is plastic cling wrap wrapped over construction paper. This allows the fluid drops to be visible and does not create any surface tension. If you can find another suitable surface which lets you out of this extra work, then by all means use it. Set out the small containers of each liquid and the medicine droppers on the table as well.

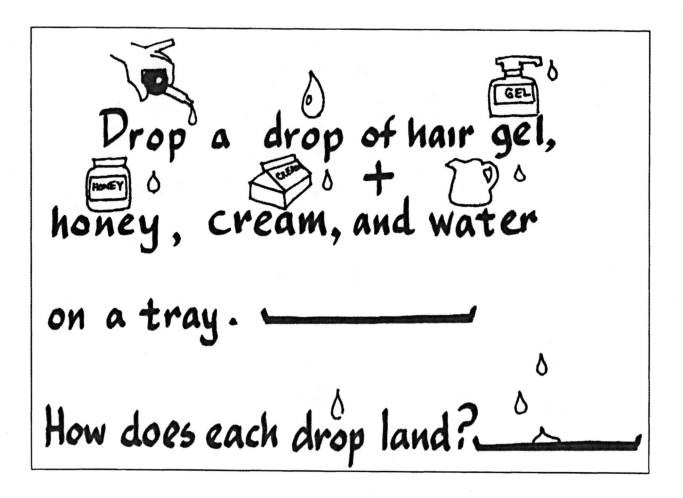

Drop a drop of hair gel, + honey, cream, and water on a tray. ———— How does each drop land?

During an Attention Getter time, show the children the materials on the table and, together, identify each liquid. Draw up one of the fluids with a medicine dropper, and hold it over some plastic wrap. Ask the children to predict whether each one will fall with a plop, in one drop, or whether it will splatter into a lot of little drops. Rather than releasing the fluid to find out, ask the children what they see in the room that will help them conduct this experiment for themselves. Facilitate experimentation. After the children have explored for a while, take your puppet out. Use your puppet to ask your students about what they are doing and what results they notice. What happens to different fluids when they are dropped from a height? What do the results of the experiment have to do with the different densities of the liquids? Also, ask the children to notice how the suction of the medicine droppers works differently with liquids of different densities.

Note: Suction is the production of a vacuum in a cavity so that external atmospheric pressure forces the surrounding fluid into the cavity. If the children ask you what suction is, one way of explaining might be: "When you squeeze the rubber part of the medicine dropper, it forces and pushes all the air out. Then when you let go of the rubber part, because there's no air in the dropper, the air outside the dropper helps to force the fluid into the tube part. That is suction."

Fluid Roll
Science

WHY we are doing this project: to help children understand the connection between the speed with which a liquid rolls down a ramp and the density of the liquid; to develop fine motor skills; to develop self-esteem and a sense of autonomy through use of a one-person work station; to develop all components of language arts: reading, writing, speaking, and listening.

WHAT we will need:

Four long wooden planks or blocks (to be used as ramps)

Support blocks

Honey

Hair gel

Colored water

Cream

Four small containers

Four eyedroppers

Newspaper

Activity sign (provided on page 120; photocopy and enlarge for your use)

"One person may be here" sign (provided on page 165; photocopy and enlarge for your use)

Tub of warm, soapy water

Paper towels

Butcher paper

Two markers

Demonstration:

Four spoons

HOW we will do it: You can set this project out either on the floor or on a table. In either case, spread several thick layers of newspaper underneath. Make four ramps that slant downward by propping up each long block or plank with several blocks. (See illustration.)

Make sure you make each ramp exactly the same. Again, put a very small amount of each fluid in a small container to avoid too much of a mess, and put an eyedropper in each one. In a prominent place, tape up the "One person may be here" sign and the activity sign. Put a small amount of warm, soapy water in a tub and place it and several paper towels near the project. These are to be used by the children to clean and dry the ramps between experiments.

If you like, use the sample prediction chart in the text to make your own on butcher paper, or create your own headings. Your container symbols should reflect the actual size and shape of containers you have. Tape the chart up near the place where you gather at Attention Getter time.

During an Attention Getter time, show the children the original hair gel container, the cream carton, a pitcher of water, and the honey bottle or jar. Together, as a group, name the fluid in each one, and then identify the corresponding symbol for that fluid on the prediction chart. Show the children the eyedroppers and ramps. Say: "Let's predict how fast or how slow each of these fluids would roll down the ramp if you used the eye droppers to put a drop of each of the fluids on a ramp." If you are working with three- and young four-year-olds, before you write down the children's predictions about each liquid, take a spoon and spoon some of each fluid out of, and back into, the container so that the children can see their relative densities. As you do this, ask: "Is this fluid a thick, dense fluid or is it thin and watery?"

With older children, just show them the liquid containers (cream carton, honey bottle, and so forth) ask about the density of each one, and elicit predictions. Ask the children to write their predictions on the chart, or do this yourself. If you print the predictions, use quotation marks and write the children's names after their comments. Use alternate marker colors to make each sentence easier to identify. Discuss the "One person may be here" sign and the purpose of the soapy water and paper towels. Show the children how to use them to clean and dry the ramps between experiments.

Invite the children, one by one, to conduct the liquid roll experiment and to make their own discoveries. Afterward, ask the children about what they found out through their experiments. Have the children read the prediction chart, or read it to them. Were there any surprises? What did they notice about which fluid rolled the slowest down its ramp? What did they notice about which one rolled the fastest? Did any of the liquids get stuck and not roll down at all? Leave the chart up on the wall for a few days.

Drop a drop of each liquid

onto a ramp.

Which rolls slowest?

Which rolls fastest?

Science Experiment Prediction Chart

Will hair gel roll quickly or slowly?

Will honey roll quickly or slowly?

Will water roll quickly or slowly?

Will cream roll quickly or slowly?

The Story of Greedy Gert

Language/Multicultural/Anti-Bias

WHY we are doing this project: to help children develop listening skills; to expand vocabulary; to help children develop their imaginations; to create multicultural, anti-bias awareness; to generate interest in the math activity immediately following.

WHAT we will need:
  Flannel board
  Flannel board shapes (provided on page 122; photocopy and enlarge for your use)
  Felt
  Glue
  Markers or crayons
  Scissors
  Clear contact paper
  Double-sided tape
  Story: "GREEDY GERT AND THE LIQUIDS" (provided)

HOW we will do it: To prepare, color in the flannel board story shapes, and glue them onto felt. Cut out the shapes, cover them with clear contact paper on the non-felt side, and put double-sided tape onto the front of the yoke shape so that it will stay attached to Greedy Gert.

If you work with very young children, tell the story more than once over a period of days or weeks and discuss a few of the new words each time. A vocabulary list follows the story.

"GREEDY GERT AND THE LIQUIDS"

Once upon a time there was a very wise Queen who ruled all the land. [Put queen shape on the flannel board.] And this Queen heard of a woman in her kingdom who was very, very greedy and because of this, people called her Greedy Gert. Greedy Gert cheated the people who bought things from her shop, and one hard winter, when food was very scarce, and the townsfolk were desperate for something to eat, Greedy Gert charged them very high prices for the food they bought just because she knew the people had

no choice but to pay it. So angrily they asked the Queen to make Greedy Gert leave the kingdom forever, but the Queen thought she would give Greedy Gert one more chance. The Queen called her to the palace. [Put Greedy Gert on the board.]

"Greedy Gert," the Queen said, "The people want you to leave the kingdom because of your greedy ways, but I'm going to give you one more chance. I'm sending you on a long journey to one end of the kingdom and back. And on this long journey, you must help everyone who asks you. All you may take with you is a yoke and two magic cups that are bottomless and can never be emptied. [Put up the yoke shape.] In these cups you may have water or liquid gold. Which will you choose?" Guess which one Greedy Gert chose! [Let children answer.]

The wise Queen said, "Gert, do you know that liquid gold is much heavier than water, and that you have a very long way to walk?"

But Greedy Gert only said, "Two whole bottomless cups of liquid gold!! I'm rich, I'm rich!! I'll keep this gold with me forever and when I get back from your stupid journey, I'll be the richest woman in town!! Ha ha hee hee hoo hoo ha!!" [Laugh greedily.] And she set off on her journey. [Stick yoke to Gert's shoulders; remove Queen.]

Well, Greedy Gert walked and walked and many times the liquid gold was so heavy that her shoulders ached, and many times she was so thirsty that she thought longingly of the water she could have had, but then she thought of how rich she would be when she got back to the town, and the thought of that made her greedy heart glad and kept her trudging along.

Well, on the road one day, Greedy Gert came across a man who was homeless and who held out a tin cup. [Put homeless man on the flannel board.] He looked dirty and tired and sick, and he said, "Oh please, ma'am, give me something to help me. I haven't eaten in a week, and I'm so hungry." Now Greedy Gert remembered that the Queen had said she should help everyone who asked her, but she thought the Queen would never know whether she really did or not, so Greedy Gert just walked by that man as fast as she could and pretended not to hear him even though it was hard to walk fast with those heavy, heavy cups

Enlarge these shapes on a photocopier.

Cover with contact paper and put double-sided tape on the yoke so that it sticks onto Greedy Gert when you move her.

of liquid gold. [Remove homeless man.] On she trudged, though carrying all that gold always made her back and shoulders and neck ache. Then one day, in another town, she met a little girl [put little girl on flannel board] who was weeping because her puppy was very sick, and she had no money to pay a veterinarian to help it. She asked Greedy Gert if she would lend her some money, but Greedy Gert walked by as fast as she could and pretended not to hear her. [Remove girl.]

Well, finally one day Greedy Gert came close to the end of the kingdom. At the end of the kingdom was a desert. [Put up desert scene.] Her shoulders and back were hurting badly because of how heavy the liquid gold was, and it was hard for her to walk through the sand. The air in the desert was very hot and dry, and as Greedy Gert trudged along she thought about how very soon now she would be back at home in the town again, and how she would be the richest person there. She was trying to decide what to keep all her liquid gold in—whether she should lock it in a safe or bury it in the ground—when she looked up and realized that she was lost!! She'd been so busy thinking Greedy Gert thoughts that she hadn't watched where she was going. She walked and walked, trying to find her way, and as she did, the sun in that desert burned more and more hotly and Greedy Gert got more and more tired and more and more thirsty until her throat ached and her mouth was as dry as a bone and she collapsed on the sand and was afraid she was going to die. "Oh, why didn't I choose the water?" she thought to herself. "I'm more thirsty than I have ever been in my life and I cannot drink this liquid gold. I'd give anything if only I could go back and choose the water instead!"

Well just then, the wise Queen floated down from the sky because she was a magic Queen. [Float Queen down onto the board.] It had really been she, disguised as the homeless man, who had asked Greedy Gert for help, and it had really been she, disguised as the weeping little girl, who had asked Greedy Gert to help her puppy. The wise Queen said, "Greedy Gert, I hear your wish, and if you have learned the lesson I intended, I will grant it. What have you learned?"

Well, Greedy Gert thought and thought and finally she said: "I never believed it before, but I guess there really are some things that are more important than gold."

"Like what?" the Queen asked, looking sternly at Greedy Gert.

"Well, when people are in pain and when they need help, I guess helping them is more important than being rich. I know that now because of what a horrible time I have had in this desert."

And the Queen said, "Yes, that is the lesson." And with her magic, she filled the bottomless cups with water instead of liquid gold and Greedy Gert drank and drank and drank. And when she stood up, the cups were so much lighter because now they were filled with water instead of liquid gold. Then, by magic, they found themselves back in the town. The Queen let Gert keep the bottomless cups of water to remind her forever of her journey, and Gert always thought it was the most delicious water she ever tasted. And Greedy Gert changed so much in the way that she treated people, that the townsfolk had to call her Gracious Gert and Generous Gert instead. And that is the end of that story.

VOCABULARY:
Kingdom
Scarce
Weeping
Folk
Collapsed
Desperate
Intended
Angrily
Sternly
Aware
Trudging
Generous

After telling and discussing the story, put the flannel board and pieces out for the children to retell the story and make up new ones. Put a "Two people may be here" sign near the materials.

Weighing Liquids

Science/Math

WHY we are doing this project: to enable children to compare the weight of a dense liquid like honey to the weight of a less dense liquid like water.

WHAT we will need:
- Balancing scale
- Small paper cups
- Small coffee scoopers
- Honey or corn syrup
- Water
- Two large containers
- Sensory table

HOW we will do it: Put the honey or corn syrup in one of the large containers and do the same with the water. Place these next to the scoopers, balancing scale, and paper cups in the sensory table.

After you tell the Greedy Gert flannel board story, say: "This is our pretend liquid gold (show them the honey or corn syrup) and this is some water." Show them the scoopers, cups, and scale and ask them to predict which one is heavier. Encourage the children to explore the materials and to verbalize their findings.

Developmental differences: Three- and young four-year-olds will not be concerned with weighing the exact same amount of each fluid, but in their sensory exploration of the materials (including the scale), they will learn, in their own way, that one fluid is heavier than the other. They may also enjoy mixing the two liquids with their fingers. Older children will enjoy scooping the liquids and will be more inclined to put both on the scale to observe what happens.

Liquid Squeeze

Science/Art/Sensory

WHY we are doing this project: to help children observe the difference in consistencies between thick and thin paint; to provide a sensory experience; to develop creative expression.

WHAT we will need:
- Paper
- Tempera paint powder
- Water
- Squeeze bottles
- Liquid starch (found in laundry supply section)
- Newspaper

Teacher:
- Funnel

HOW we will do it: Well ahead of time, start collecting small squeeze bottles like honey containers, small mustard or ketchup squeeze bottles, and lotion dispensers. To begin the activity, spread several layers of newspaper over your work area. Mix up many colors of tempera paints, being sure to include black and different shades of brown. Vary the consistency of the paints by adding more or less water and liquid starch. Some paints should be very watery and others should be fairly thick. Put *small* amounts of paints in the squeeze bottles and use the funnel to refill as necessary. Set the bottles and the paper on the table. Encourage the children to make squeeze paint pictures by squeezing paint onto paper and then using their fingers to spread the paint. Talk about how quickly or slowly the different paints flow out of the squeeze bottles and why.

Marble Roll

Science/Art

WHY we are doing this activity: to allow children to observe what happens to marbles rolling through thin and thick paint; to develop hand–eye coordination; to develop creative expression.

WHAT we will need:
    Shirt boxes
    Tempera paint powder
    Paper
    Spoons
    Marbles
    Water
    Liquid starch
    Small containers
    Newspapers

HOW we will do it: Spread several layers of newspaper over your work surface. Mix your tempera paint with less or more water and liquid starch so that you have thick, pasty paints as well as more watery ones. For a medium consistency, add tempera powder directly to liquid starch. Put a small amount of paint into each container. Put a piece of paper in each shirt box, and put all the boxes on the activity table with the paint, marbles, and spoons. Invite the children to spoon paint into a box and roll marbles around. What do they notice about how the marbles roll through the thick paint? What happens when the marbles roll through watery paint? Compare the tracks made by both.

River, River

Music/Movement/Gross Motor/Language/Cognitive/ Small Group Activity

WHY we are doing this project: to reinforce, through music and movement, the difference between the movement of dense liquid and watery liquid; to develop the large muscle group; to encourage children to express themselves creatively through physical movement; to help children feel comfortable using their singing voices; to develop cognition through memorization of song words.

WHAT we will need:
    Butcher paper
    Markers
    Song: "RIVER, RIVER" (to the tune of
      "Oh My Darling Clementine")
      (provided)

HOW we will do it: Pin up your butcher paper near the place where you gather for Attention Getter time. When the children are gathered, tell the following story, and use the markers to draw what you talk about. (Do not feel self-conscious about your pictures—I am no artist, and my drawings are always as rough as they can be, but the children still enjoy them.)

Before you begin, ask the children: "What kind of liquid do you find in a river?" When they answer, "Water," tell the story:

"THE RIVER"

Once upon a time there was a stream, and it flowed past trees and mountains and even a town. But one day the stream got very bored and decided to flow somewhere different. It met up with a mighty, rushing river that roared and tumbled its way along.

"May I join you?" the stream asked.

"As you like, as you like, but hurry along, no time to stop," the river said. So the stream joined the roaring river, but oh dear, oh dear, it didn't like it at all! They rushed and roared and tumbled along so fast that the stream had no time to look at what they were passing, and it was altogether too much. Well, as it travelled along with the huge, roaring, tumbling river, the stream saw a little creek branching off to the side. It said, "Little creek, may I join you?"

Well, the little creek said in a slow, lazy voice: "As you like, as you like, we have got all day and all night too." So the little stream branched off and joined the creek. Well, at first it was such a wonderful change not to be roaring and tumbling along all in a rush. But soon the little stream became very bored because they trickled along sooo very slowly and sooo very sluggishly and day after day there was nothing new to see and only the same old things to do. And, worst of all, the little stream noticed that it really wasn't a stream anymore. It was getting thinner and thinner until it was really just a little muddy dribble of water.

Well, luckily that night a big rain came. It rained and rained for hours and hours until it swelled up all the waterways of the land. The little stream was so swollen now that it burst the banks of the tiny creek, and branched away down a valley. And the little stream was very happy to be its old self again, splashing and gushing over rocks and stones, forming little falls and pools, and chattering and babbling away to itself under the hot, bright sun. And that is the end of that story.

After you tell the story, say: "Now we are going to be a river." Have the children line up and put their hands lightly on the shoulders of the person in front of them. Wind and twist your way around the furniture in the room as you sing this song:

"RIVER, RIVER"

Roaring river, roaring river, roaring river, roaring free,
River rushing and a-gushing,
streaming down to meet the sea.

After the children learn the words, sing together and wind your way around the room as quickly as you can without tripping, falling, or bumping into furniture. Then say to the children: "What would a river of honey move like?" Discuss this for a while, and then suggest that you all pretend to be a river of honey. Explain that *sluggish* means slowly and lazily. Ask the children if they have ever seen a slug. How did it move? When you are ready to be a river of honey, wind your way sluggishly around the room and sing very sluggishly:

Sluggish river, sluggish river,
See it oozing sluggishly,
See it creeping and a-crawling,
slowly rolling to the sea.

Dish Wash
Science/Dramatic Play/Language

WHY we are doing this project: to help children learn to express their ideas; to develop imagination; to facilitate social interaction with other children; to help children coordinate actions with words; to enable children to compare the density of liquid dish soap to water.

WHAT we will need:
    Dramatic play kitchen furniture
    Sensory table or basins
    Play dishes
    Miniature liquid soap bottles (if possible)
    Liquid soap
    Funnels
    Strainers
    Pitcher of water

HOW we will do it: Miniature liquid soap bottles containing sample portions are sometimes available at variety stores. If you cannot obtain these, put your liquid soap in the standard-sized bottles. Spread several layers of newspaper under your sensory table or basins, fill them half or one-third full of warm water, and place them in your dramatic play kitchen area. As the children explore the materials, invite them to compare what happens when they squeeze liquid soap into a strainer to what happens when they pour water through it. How does each liquid make its way through the mesh? Invite the children to notice the speed with which each liquid flows. As the soap is mixed with the water, does it change the texture or density of the water? Discuss!

Making Lassi

Multicultural/Math

WHY we are doing this project: to develop an appreciation for the cuisine of other cultures; to enable children to make a recipe with three liquids of different consistencies; to familiarize children with cup and spoon measurements.

WHAT we will need:

Book: Kalman, Bobbie, *India the Culture* or
    Ganeri, Anita, *Journey Through India*
Plain, nonfat yogurt (one cup for
    each child)
Honey
Cinnamon
Cold water
Four bowls
1 cup measuring cup
⅓ cup measuring cup
Two teaspoons
One ½ teaspoon
Permanent marker
Blender
Clear plastic cups (large enough for
    1½ cups of liquid)

HOW we will do it: Set out everything on a table except the books, marker, and blender. Put the yogurt, water, honey, and cinnamon in separate bowls. Using the permanent marker, mark the measurements of the cups in clear, large numbers and letters. Put the 1 cup measuring cup in front of the yogurt, the ⅓ cup in front of the cold water, a teaspoon and the ½ teaspoon in front of the honey, and a teaspoon in front of the cinnamon.

During an Attention Getter time, look through the photographs in *India the Culture* or *Journey Through India* and talk about them together. Tell the children that it can get very, very hot in some parts of India, and that people like to drink a very cool, refreshing drink called *lassi*. Say this word together several times. Invite the children to each take a clear, plastic cup and to fill it by taking only *one* spoon or cup of each spice or liquid. If you like, walk around the table, pick up each measuring utensil, and go over this with the children before they begin.

Invite the children to put all ingredients in their cups, and as they do, ask them what they notice about how the yogurt, honey, and water are different. What did they notice about how easy or hard it was to get each fluid out of its measuring spoon or cup? Have the children bring their ingredients to you, mix them in the blender, and pour the *lassi* back into their cups to drink. You may want to sweeten your *lassi* more than is called for in this recipe. Make it to your and the children's taste.

Literature

Symbol Key: *Multicultural
    +Minimal diversity
    No symbol: no diversity or no people

Allen, P. (1980). *Mr. Archimedes' bath.* Lothrop, Lee & Shepard Books.

Ardley, N. (1991). *The science book of water.* San Diego, CA: Harcourt Brace Jovanovich.

Burns, D. (1990). *Sugaring season.* Minneapolis, MN: Carolrhoda Books. (This book has some excellent photographs of how maple syrup is made.)

Ganeri, A. (1994). *Journey through India.* Mahwah, NJ: Troll Associates.*

Halpern, S. (1992). *My river.* New York: Macmillan.* (Beautiful color pictures.)

Kalman, B. (1990). *India the culture*. New York: Crabtree.*

Ravzon, M. J., & Overbeck Bix, C. (1994). *Water, water everywhere*. San Francisco, CA: Sierra Club Books.* (This book has excellent photographs and does a good job of showing the effects of acid rain and water pollution.)

Turner, D. (1989). *Milk*. Minneapolis, MN: Carolrhoda Books.

Extenders

Science: Provide fluids of different densities, including hair gel. Invite the children to pour some of each into a paper cup. In which liquid does a Popsicle stick stand straight up? Why?

Science: When you conduct your liquid roll experiment down wooden ramps, let a drop of four liquids of different densities each fall on a ramp at the same time, and let them race each other. Watch as a group, and as the liquids are rolling, ask the children to predict in what order the drops will reach the end of the ramp.

Movement/Gross Motor: When you sing the "RIVER, RIVER" song, vary your movements by rolling and tumbling instead of walking. Spread gym mats out on a large open area. Have the children roll or somersault as a tumbling river, and then as a sluggish river.

The Five Senses

Attention Getter: Cut several oranges into quarters, so that you have one quarter for each child. Have these on hand, as well as paper towels and small paper cups. Pass a cup and one orange quarter to each child. First, ask the children to lay the orange piece on the paper towel, and to look at it close-up. Ask them to pretend that they have never seen an orange before, and do not know what it is. Then ask them to close their eyes, and touch it with their fingertips. How does it feel? Next, ask everyone to squeeze the orange piece into a paper cup, and to listen careful-ly. Does the orange piece make a sound when it is squeezed? Ask them to close their eyes and smell the juice. Ask: "If you didn't know what was in your cup, could you tell that it's orange juice?" Last, ask everyone to drink their juice and to concentrate on the taste on their tongues.

When you have followed all the above steps, ask the children what parts of their bodies they had to use for these activities. After the children express their ideas, point to the appropriate part of your body as you mention it and say: "We used our sense of sight, our sense of hearing, our sense of taste, our sense of smell, and our sense of touch. We just used all five senses." Go back over each step you took in your discovery of the oranges, and ask the children what they remember about which senses they used with each step. Ask the children to guess what you will be talking about and working with over the next few days.

Taste: Taste Center
Science/Sensory/Language

WHY we are doing this project: to help children become aware of the sense of taste; to facilitate a sensory experience; to develop speaking and listening skills; to develop cognition by helping children learn the difference between sweet, sour, salty, and bitter; to help children understand that certain parts of the tongue are more sensitive to certain tastes.

WHAT we will need:
Plates
Slices of pickle
Potato chips
Lemon slices (very small)
Chocolate chips
Small paper plates
Activity sign (provided on page 131; photocopy and enlarge for your use)
Butcher paper
Tongue diagram (provided on page 131; photocopy and enlarge for your use)
Markers
Puppet
Older children:
Toothpicks

HOW we will do it: Cut the pickles into slices and slice the lemon into very small pieces. Put all the food out on plates, set the plates on the activity table, and put the toothpicks beside them. Make the activity sign according to the suggested format in this text, and on the butcher paper, make an enlarged chart of the tongue diagram in the text.

During an Attention Getter time, point to all the foods and ask the children to predict how each will taste. Next, ask them to predict whether parts of their tongues will be able to taste some of the foods better than other parts of their tongues. Let the children know that they can conduct a scientific taste experiment. Reinforce the part of the activity sign that says: "Please eat only one at a time." Ask: "What would happen if one person took a whole bunch of chocolate chips and ate them?" (No one else would get to try the experiment.) Also, make sure the children understand that they can put things on their paper plates, carry them back to the table, and that they must sit down while they eat.

Let the children know that it is easier to tell what parts of the tongue are tasting if they just eat a little piece of something. Show them the toothpicks, and encourage them to scrape a little bit of food onto a toothpick, and then to touch it onto different places on their tongues. If you are working with very young children, do not include the toothpicks in the activity, and encourage them to use the tips of their fingers instead. Also, if you work with three- and young four-year-olds, you will probably have to supervise this activity fairly closely. After the children have had a chance to conduct the experiment, ask them about their findings. Ask them if the lemon piece and pickle slice are sour in the same way.

When appropriate, take out your puppet and have it conduct the experiment and make wrong predictions. For example, the puppet says to one of the children: "Oooh, I'm going to try a chocolate chip. Boy, I bet it's going to be sour." Or: "I think I feel like eating something sweet now. Hand me a potato chip."

After the experiment, read the tongue chart to interested children. Did it turn out to be accurate?

130

Take a taste of
pickle, potato chip, lemon slice,
and chocolate chip.
1
Please eat one at a time!

Smell: Smelling Bottles
Science/Sensory/**Optional:** Language

WHY we are doing this project: to help children develop awareness of the sense of smell. **Optional:** to develop reading skills.

WHAT we will need:
- Plastic spice bottles
- Glue
- Cloves
- Allspice
- Cinnamon
- Coffee
- Rubbing alcohol
- Paper
- Perfume or essence oil

HOW we will do it: Dip a piece of paper in rubbing alcohol, put it inside a spice bottle, and glue the perforated lid on. If the bottle has no perforated lid, glue the regular lid on and then poke holes in it with a skewer. Repeat this process with perfume or essence oil. Make smelling bottles in the same way with the rest of the materials by putting some of each item inside a bottle. The above list is really just to give you ideas; there may be other things you would like to put in your smelling bottles.

Depending on how much time you have for preparation, you may decide to make this a language development activity also. Write on sheets of paper the name of each object inside each bottle and glue a sample next to it. For rubbing alcohol and perfume, draw a picture of the original bottles beside the word. Put your activity table against a wall, and tape each sign onto the wall. Place each smelling bottle in front of its corresponding sign. Lay out blank sheets of paper, pens, and crayons for the children to scribble, write, or draw pictures. Some children may use the writing materials to write about or draw the smelling bottles; other children may use the materials to express other ideas.

Set all the bottles on an activity table and invite the children to explore. Encourage them to describe what they smell. Do they have a

favorite smell? Which is their least favorite? Another variation of this is to put familiar-smelling things inside the bottles, cover the bottles so their contents cannot be seen, and encourage the children to guess what is inside them. Some ideas include: mint candy, talcum powder, a piece of orange, perfume, toothpaste, pine cone.

Smell: Making Potpourri (Part 1)
Science/Crafts/Gross Motor

WHY we are doing this: to help children become aware of the sense of smell; to help children enjoy making crafts; to develop the large muscle group; to develop speaking and listening skills.

WHAT we will need:
- Commercial potpourri (small amount, the kind that consists of pure botanicals, no fillers)
- Sandwich bags
- Whole cloves
- Cinnamon sticks
- Vanilla beans
- Whole allspice
- Butcher paper
- Marker
- White labels
- Clear plastic salad containers

HOW we will do it: To obtain your plastic containers, go to a deli and ask if they will donate some of these containers, or if they will sell you some; they are not expensive. The reason that these containers are ideal is because they are safe (not glass), they are transparent, and they have lids. If you come across other containers that meet these criteria, then use those. When you get your plastic containers, prepare them for the activity by using the skewer to poke several holes in the lid of each one, or if this is too difficult, cut small slits with an Exacto knife.

This allows the scent of the potpourri to be smelled even when the container is closed.

Cloves, cinnamon sticks, vanilla beans, and allspice are expensive. There are a few ways you can get the amount you need for all your children without breaking the bank. If you work for a nonprofit school, visit your local supermarket, explain your needs, and ask the store manager if she or he is willing to donate some spices. If you have a school newsletter that goes out to parents, tell the manager that you will thank the store in the newsletter (free advertising). Another option, if you are a teacher, is to ask each parent to buy one container of one of the spices.

To prepare, tape up your piece of butcher paper on the wall near the place where you gather for Attention Getter time. Write each child's name on two white labels. Apply one to the sandwich bags and reserve one for the plastic containers. During an Attention Getter time, show the children your commercial potpourri. Present it in a pretty container. Pass it around so the children can smell it and touch it. Explain to the children that it is called *potpourri* (the first syllable rhymes with "go": PO-PURR-EE) and print the word on the butcher paper. Explain that it is a French word that originally meant "stew." What is a stew? Why is the potpourri like a stew? Discuss this and say the word several times as a group. Explain that you are going to make your own potpourri, and that you will have to go out looking for things with which to make it. Go on a nature walk, and depending on the time of the year and where you are located, try to go some place where the children will be able to pick up pine cones, eucalyptus leaves, and flowers or flowering weeds. Before you set out on your walk, give each child his or her own sandwich bag. Bring home your materials and put them out to dry. An easy way to keep them separate is to have each child lay his or her materials on top of his or her sandwich bag.

When the materials are dry, have the children select cinnamon sticks, allspice, vanilla beans, and cloves to mix with the ingredients they found on the nature walk. Have the plastic containers available on the table so that each child can choose one to contain the potpourri. If you like, have the name labels on the containers already, and invite the children to find the one

that has her or his name on it. At the end of the day or the session, encourage the children to pass their potpourris around to experience the scent of the different mixtures they made.

Touch: Letter and Number Boards

Science/Sensory/Language/Math/Fine Motor

WHY we are doing this project: to help children become aware of the sense of touch; to develop reading and writing skills; to practice rational counting; for kindergartners, to facilitate addition; to develop fine motor skills.

WHAT we will need:
  Sandpaper
  Stencils (provided on page 134;
    photocopy and enlarge for your use)
  Cardboard
  Small uncooked pasta shapes (or small
    beads, beans, or spangles)
  Glue sticks

HOW we will do it: In this project, you're going to make cards consisting of a numeral, as well as the numeral spelled out in sandpaper letters. Children can touch and see the word and number.

To prepare, cut out the provided stencil shapes, and use them to cut out sandpaper letters and numbers. You can ask friends, family, neighbors, and other parents to help you with this several weeks in advance. Cut out cardboard rectangles about 20 cm x 10 cm (8" x 4"). Use a glue stick to attach a sandpaper word and numeral to each one. If you work with three- and young four-year-olds, also glue on a corresponding number of sandpaper circles. If you work with older children, this step is probably not necessary, though you can make that judgment yourself.

For kindergartners, make addition cards. Example: "Two and three are ___." Use the stencils to make sandpaper words and num-

bers for "two" and "three." On an activity table, set out the number cards, glue sticks, and containers of small beads, pasta shapes, or beans. Also, put out blank cardboard rectangles, sandpaper letters and numbers, and uncut sandpaper sheets for children who want to make their own number cards from scratch. If you are using glue sticks, the objects the children will attach must be small (spangles are perfect). If you plan on using liquid glue and brushes, the objects can be bigger.

Ahead of time, take a numeral card, and glue on the number of objects it specifies. During an Attention Getter time, say: "I'm going to pass something around, and I want you to touch it with your eyes closed. See if you can feel which number it is, and if you know, don't say it out loud. Keep it a secret in your head." Pass the card around and when everyone has had a chance to feel it, see what number the children think it is. Show them how you glued on that many beads (or whatever). As a

group, count the sandpaper circles on the card (which should correspond to the numeral) and then count the beads. Ask the children what they see in the room that would help them make their own number cards.

If you are working with kindergartners, comment on their addition problems as they work them out.

Developmental differences: Three- and young four-year-olds are likely to make the gluing of objects an end in itself, and will probably not match numerals with the number of objects they put on their cards. They will be using their sense of touch as they use the sandpaper, small objects, and glue. If you want to make this a math activity also, invite each child to count the total number of things glued onto the tagboard when the child is finished. Older children will be more interested in feeling the sandpaper numbers and letters, counting the number of objects designated on the cards, and also, in making their own.

134

Touch: Feely Box Center (Part 1)

Science/Sensory/Language

WHY we are doing this project: to help children become aware of the sense of touch.

WHAT we will need:
>Eight cylindrical oatmeal containers
>Eight knee-high stockings or socks
>Piece of velvet
>Sandpaper
>Styrofoam pieces
>Gravel
>Marbles
>Plastic bubble packing
>Wet sponge
>Plastic sandwich bag
>Two eggs
>Glue

HOW we will do it: Place one of each object inside an oatmeal container. Cut the stockings off at the ankles, and then stretch one end of a stocking over the opening of each oatmeal container so that the tube extends beyond the opening. Socks and nylons are usually tight enough to stretch across the container securely. The children will pull the stocking over their hand in order to reach inside the container. To make the egg feely box, crack open both eggs and slide the contents into a baggie. Seal the top by putting glue along the seam before you press it closed. Then put the sandwich bag in the container.

Set all these containers on an activity table and encourage the children to reach inside them and feel the contents. Can they guess what they are?

Touch: Feely Box Center (Part 2)

Science/Sensory/Language

WHY we are doing this project: to use the sense of touch in developing all aspects of language arts: speaking, listening, reading, and writing.

WHAT we will need:
>Feely boxes (from previous activity)
>Paper
>Markers
>Crayons
>Pens
>Word cards (instructions follow)
>Magazines

HOW we will do it: To prepare, make the cards by using a black marker to write words in bold, clear print: soft, rough, cold, gritty, spongy, squishy, damp, and any other words which describe the objects in the feely boxes. Cut out magazine photographs of things which can be described by one of these adjectives. Glue the picture next to the appropriate word, or, if you cannot find suitable photographs, draw a picture.

Make a writing center near the feely boxes. Set out markers, crayons, pens, blank paper, magazine photographs torn out of magazines, scissors, word cards, and glue sticks. On the wall by the writing center, tape up some of the word/picture cards you made in the paragraph above.

The materials can be explored in many variations, depending on the interests and level of language development of your children. Children can draw a picture of the object they think is in a feely box. You can take dictation of what the children felt, spell words for the children, support invented spelling or scribbling, or write down their words to be copied onto their pictures. Children can find a magazine photograph which corresponds to one of the words on the word cards, and glue the picture onto the card, or make a whole new card. Or they can glue a magazine picture onto paper and write or give story dictation which describes how it might feel. Or, if you and your children think of other ways to use the materials, go with it!

Touch: Texture Art
Art/Science/Sensory

WHY we are doing this project: to facilitate creative expression; to develop awareness of the sense of touch; to give children the opportunity to experiment with unusual art materials.

WHAT we will need:
Flour
Corn meal
Aquarium gravel
Coffee grounds
Ground black pepper
Cinnamon powder
Glitter or colored sand
Glue
Squeeze bottles
Paper
Newspaper
Small containers
Tub with small amount of
    soapy water
Paper towels
Optional:
Tempera paint powder

HOW we will do it: Spread newspaper out on your activity table and put all the materials in the small containers. Put glue in your squeeze bottles, and if you like, thin it out by adding a little water. For interest, add tempera powder or food coloring to the glue as well. Put the tub of soapy water and the paper towels nearby for easy hand washing. Encourage the children to squeeze glue onto paper, and to use their fingers to sprinkle the different materials on the glue. How does each one feel? Introduce descriptive words into the conversation if necessary (e.g., *gritty*, *soft*, *coarse*), as the children make their pictures. Feel free to add other differ-

ently textured materials to this activity to make it all the more interesting. Ask the children to notice how the coffee grounds, cinnamon, and pepper smell as they use these materials.

Hearing: The Tape Recorder Game
Science/Sensory/Cognitive/Small Group Activity

WHY we are doing this project: to help children become aware of the sense of hearing; to develop competence with technology; to develop cognition through recognition of each other's voices.

WHAT we will need:
Tape recorder
Tape

HOW we will do it: This is a small group activity. Have one child leave the room (to a place which is supervised by an adult or where you can still see the child) or simply have the child go to a far place in the room, away from the group. Record one child's voice, and have the first child return. Play back the recording. Can the child guess whose voice it is? Let the children take turns with both roles, and also, show them how to operate the tape recorder. Show them the symbols on the buttons that represent *record*, *rewind*, and *play*. Let your students take turns operating the machine.

Hearing: Sounds in the World Around Us

Science/Social Studies/Small Group Project

WHY we are doing this project: to develop appreciation for nature; to make children aware of the sense of hearing; to create a restful, relaxing mood for quiet time.

WHAT we will need:
> Tapes of natural sounds (e.g., ocean, rain forest, birds, whale, or dolphin sounds)
> Tape recorder
> Paper
> Crayons

HOW we will do it: This is another group activity. Many libraries now carry tapes of natural, environmental sounds so you may want to try this as a source. Also, a recording studio called "Moods" at Gateway Recordings, P.O. Box 5087, FDR Station, N.Y., N.Y., 10022 has a wide variety of environmental tapes including "Backyard Stream," "Farm Life," "Metropolis," and "Thunderstorm Terror." You may want to send for their brochure and invest in a few tapes. Many variety stores also carry these kinds of tapes.

Gather your children together and dim the lights to create a relaxing environment. Play many different sounds from the tapes that you have, and ask the children to guess what they are hearing. In between tapes and guesses, discuss what you are hearing by asking: "Have you ever heard that noise before? Where were you? What did you see?" Provide paper, crayons, and markers and invite the children to draw what they are hearing.

Sight: Tricky Pictures

Science/Sensory/Cognitive

WHY we are doing this project: to encourage children to use their eyes in a new way; to develop awareness of the sense of sight; to develop cognition; to encourage child-to-child interaction by creating a cozy space.

WHAT we will need:
Preparation:
> Construction paper
> Hole puncher
> Clear contact paper
> Magazines
> Scissors
> Large pillows or cushions

Optional:
> Large box

HOW we will do it: To prepare, leaf through the magazines and cut out large photographs of interesting objects. Fold over the construction paper pieces to make cards. Glue a photograph on the inside of each. On the covers, write: "Can you guess what it is?" Cover your picture cards on both sides with clear contact paper. Then, using a hole puncher or scissors, cut a small circle in the cover so that only a small part of the object is visible. Make several of these picture cards.

Create a cozy space for a small number of children with the cushions or pillows. If you have a large box, you can cut a door in it, and create a cozy space inside. Otherwise, put the tricky pictures on a small box or table. During an Attention Getter time, demonstrate one of the tricky pictures for the children, and then encourage them to investigate the others.

Sight: Story Seeing, Story Telling

Sensory/Language

WHY we are doing this project: to encourage children to use the sense of sight; to facilitate cooperation and interaction between children; to develop reading, speaking, and listening skills.

WHAT we will need:

Same cozy space created in previous project

Wordless picture books, such as:

Carle, Eric, *Do You Want to Be My Friend*

Tana Hoban's photographic books

Mercer Mayer's "Frog" picture books

Cardboard

Activity cards (format provided below; photocopy and enlarge for your use)

Cardboard box (food packaging: 17.5–20 cm x 10 cm [7"–8" x 4"]

Contact paper (clear and colored)

"Two people may be here" sign (provided on page 175; photocopy and enlarge for your use)

Preparation:

Scissors

HOW we will do it: To prepare, photocopy both of the activity cards provided. Using glue or double-sided tape, affix them to cardboard, and cover on both sides with clear contact paper. Cut off one end of the box, and half of the front. (See activity sign for illustration.)

Cover the box with colored contact paper. Put the cards in it, and set it out near the cozy

reading area. Pin up the activity sign and the "Two people may be here" sign nearby. Then put your wordless books in the reading area as well. During an Attention Getter time, show the children the sign and the cards and read or interpret them together. Ask the children which sense they will have to use to read the books. Invite your students to explore the materials.

Sight: The Optical Illusion of the Crescent Trick
Science/Sensory

WHY we are doing this project: to help children understand that things are not always what they appear to be; to develop awareness of the sense of sight.

WHAT we will need:

Paper (*bright* green, yellow, and red)
Crescent pattern (provided below; photocopy and enlarge for your use)
Cardboard
Glue or double-sided tape
Tray
"One person may be here" sign (provided on page 165; photocopy and enlarge for your use)

Preparation:

Scissors

The facts of the matter: When the same-size crescents are arranged in this order—red, green, yellow—the green one appears to be wider than the others. When they are arranged in this order—green, red, yellow—the green one appears to be taller. The curves of the crescents interfere with the way the mind makes sense of them, and this makes the crescents appear to be different sizes.

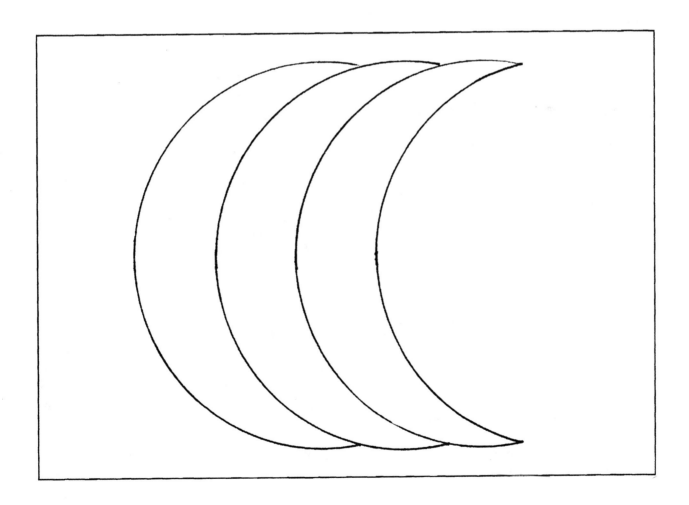

HOW we will do it: Use the crescent pattern to cut out three crescents, one from each color, all exactly the same size and shape. Using the glue or double-sided tape, affix the crescents to cardboard and cut them out; this makes them sturdier. Make the "One person may be here" sign and tape it above the area where the tray will be located.

During an Attention Getter time, stack the crescents on top of each other so that the children can clearly see they are really all the same size. Then turn the crescents upside down so that the colors are not visible and align them next to each other, so that each crescent is cradled by the curve of the next one. (See illustration, page 139.)

Ask the children to predict if, when they arrange the crescents so the colors are visible, the crescents will all look the same size and shape. Read/interpret the "One person may be here" sign together and invite the children to conduct the experiment for themselves. What do they notice?

Science kits: Make several sets of the colored crescents, and put each set in a manila envelope. Set out the envelopes for independent exploration.

Senses Working Together

Science/Sensory/Group Activity

WHY we are doing this project: to help children understand how the senses of sight, smell, and taste work together.

WHAT we will need:
    Juice (apple, orange, grape, and
        pineapple)
    For kindergartners: white grape juice
    Small paper cups
Optional:
    Blindfold

HOW we will do it: This is a group activity for Attention Getter time. Ask the children to predict whether or not they will know what kind of juice they are drinking, when they cannot see

it. Have your students take turns tasting the juices when their eyes are covered. (Either blindfold the children, ask them to close their eyes, or cover their eyes with your hand.) Can they guess which juice it is? Let the children also taste the juices by looking at them and smelling them first. Is it easier to guess correctly? Why?

Senses Working Together: The Shaking Experiment

Science/Sensory

WHY we are doing this experiment: to help children understand how the senses of touch and hearing work together; to develop listening skills; to familiarize children with a scientific process that includes: noticing a problem, making predictions, making discoveries, comparing results, and discussing them.

WHAT we will need:
    Four small coffee cans (or other tins—all
        the same size with lids)
    Gravel
    Glue
    Black marker
    Butcher paper
    Two prediction charts (format provided
        on page 141)
    Markers
    Crayons

HOW we will do it: Fill two tins with a small amount of gravel, and fill the remaining two with a large amount. Label each tin with the numbers one through four. Make two copies of the prediction chart provided or create your own, and pin them up side by side. Write a heading on one that says: "Listening Only." On the other, write a heading that says: "Listening and Shaking."

During an Attention Getter time, show the children the coffee cans, and ask them to predict whether or not they will be able to guess

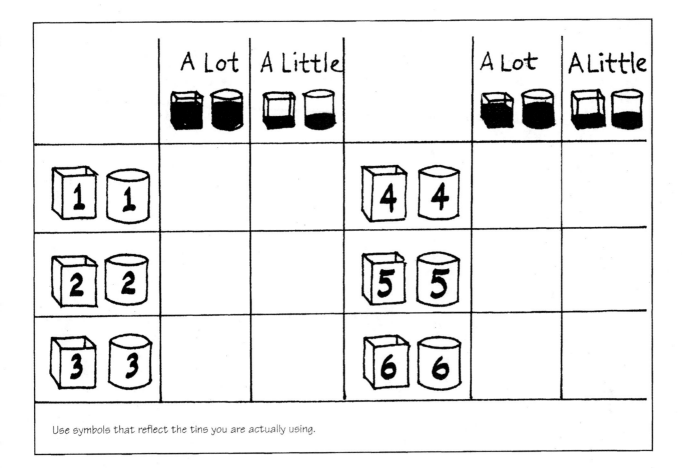

	A Lot	A Little		A Lot	A Little
1 1			4 4		
2 2			5 5		
3 3			6 6		

Use symbols that reflect the tins you are actually using.

how much is in each one only by listening to it being shaken. Have four children stand up in front of the group, and give each one a tin. Before you begin, tell the children holding the cans to zip or button their lips together, and not to tell the rest of the group whether the tin feels heavy or light, no matter how the group guesses. Have each standing child take a turn at shaking his or her can, and have the group listen carefully. After each one, have the group guess whether it has a lot of something in it, or a little, based solely on what they hear. Write down the guesses on the "Listening Only" chart by writing a check mark under the appropriate symbol for each can.

Next, pass each one around, and let each child shake it as well as listen to it. Now how do the children guess? Do the guesses for each tin change when the children can feel their weight as well as hear the contents? Write down the guesses again on the second prediction chart and compare them to the first guesses. Which was better: just hearing the cans being shaken, or hearing them shake while also

holding them? Open up the tins and let the children examine the contents. If you work with very young children, glue the lids shut. The next day, when the glue is dried, put the tins out for more exploration. Leave the charts up for several days.

What Is Braille?
Anti-Bias

WHY we are doing this project: to help children appreciate all methods of communication and to help children understand what braille is and who invented it.

WHAT we will need:
Book: Hobart Alexander, Sally, *Mom Can't See Me*
Children's books in braille

HOW we will do it: Most libraries carry books in braille, and anyone with a library card can check them out. There are also books which have clear, plastic inlay over the text, on which words are written in braille. Visit your local library, and check out several braille books. If you get books which are in braille only, try to get the same story in non-braille, so that your students know which story the braille book is telling. During an Attention Getter time, read *Mom Can't See Me* and talk about the braille materials in the photographs. Pass around the braille library books. As the children feel the raised bumps on the paper, explain that the bumps are arranged in patterns to form words. A man named Louis Braille invented this system and blind people all over the world use it. Ask: "Why doesn't the braille book have any pictures?" Read your non-braille book and while you do, let the children continue to pass around and touch the braille version. Afterward, put the books out on a table for the children to explore. You many want to talk about treating the books carefully while you are borrowing them.

What Do You Do with a Sour Lemon?

Music/Movement/Cognitive

WHY we are doing this: to help children feel comfortable with their singing voices; to develop cognition by: remembering movements associated with words, remembering each part of the song that went before, and connecting body parts with senses.

WHAT we will need:
  Song: "WHAT DO YOU DO WITH A SOUR LEMON?" (to the tune of *What Do You Do with a Drunken Sailor?*)

"WHAT DO YOU DO WITH A SOUR LEMON?"
What do you do with a sour lemon?
What do you do, oh what do you do?
Oh what do you do with a sour lemon,
I ask what do you do?

How we will do it: Sing the song with gestures. Throw your hands apart in a questioning gesture for the first three lines. For: "I ask what do you do?" point to yourself and then to the children. To answer the question, speak the line: "Smell it with your ___ (nose)." (Let the children fill in the word.) Point to your nose as you speak.

With gestures, sing the song again, follow with the spoken line, and add another: "And taste it with your ___ (tongue)." Point to your tongue as you say the line.

Repeat the song three more times, following with the previous spoken lines, and adding a new one each time: "And look at it with your ___ (eyes)." "And listen to it squirt with your ___ (ears)." "And touch it with your ___ (fingers)." Point to the appropriate body part during each line.

See if the children (and you!) can remember each sense in the list in the same order each time.

Restaurant

Dramatic Play/Language/Sensory

WHY we are doing this project: to promote child-to-child interaction; to develop speaking skills; to expand vocabulary; to facilitate imaginative, pretend play; to enable children to act out real-life situations and to work through emotions; to allow children to use all of their senses during dramatic play.

Note: The following is a list of suggestions. The more props you have to simulate a real restaurant, the more fun it will be for the children, but you may end up using very different props than the ones on this list. Use whatever

142

you have available to you, but remember to create a print-rich environment and to provide materials that utilize all the senses.

WHAT we will need:

Large boxes (to be turned over for tables)
White butcher paper (for tablecloths)
Vases and artificial flowers
Chairs
For play kitchen—utensils that make noise:
   Whisks
   Pots
   Pans
   Wooden spoons
   Chef hats
   Children's aprons
   Plastic dishes and silverware
   Plastic trays
For menus:
   Construction paper
   Food magazines
   Scissors
   Markers
For front desk:
   Notebook (for reservations)
   Carbon paper
   Pen
   Cash register
   Play telephone
Food:
   Oranges cut in quarters
   Pickles
   Potato chips
   Paper napkins

HOW we will do it: To prepare, turn your cardboard boxes upside down and cover them with white butcher paper to simulate table cloths. Arrange chairs beside them. Arrange the materials to simulate all parts of a restaurant: hostess/host station/front desk, kitchen, and dining area.

Put paper, pens, cash register, and telephone at the front desk. Put the pots, pans, spoons, and whisks in the kitchen area. Add other cooking utensils that will create the sound of food preparation.

Make menus by cutting out pictures from food magazines, gluing them onto construction paper, and folding the paper over vertically. If you can, find and use photographs of the food you are actually using. Set a menu on each table. If you like, have the children make menus as a language and art activity.

During an Attention Getter time, ask the children how many of them have been in a restaurant. Ask: "Do you remember what you heard? (Music, kitchen noises, people talking) Smelled? (Food cooking, food being served) Saw? (Everything in a restaurant) Touched? (Silverware, seats, tablecloth) Tasted? (Food, drinks)." Talk about this together. Ask the children if they see anything in the room which they could use to play "restaurant." If you decide to use real food like oranges, pickles, and potato chips for a taste experience, have the children wash their hands before they begin to explore the materials, and let them know that they must sit down while they eat.

Literature

Symbol Key: *Multicultural
+Minimal diversity
No symbol: no diversity or no people

Aliki. (1962). *My five senses*. New York: Harper Collins.*

Hoban, T. (1984). *Is it rough? Is it smooth? Is it shiny?* New York: Greenwillow Books.

Hobart Alexander, S. (1990). *Mom can't see me.* New York: Macmillan.

Your five senses. (1984). Chicago: Children's Press.*

For young preschoolers:

Oxenbury, H. (1985). *I hear*. New York: Random House.

Oxenbury, H. (1985). *I see*. New York: Random House.

Oxenbury, H. (1985). *I touch*. New York: Random House.

For restaurant dramatic play:

Barbour, K. (1987). *Little Nino's pizzeria*. New York: Harcourt Brace Jovanovich. (There are large groups of people in the illustrations who are all white. Buy the paperback edition and shade in diverse skin colors.)

Day, A. (1988). *Frank and Ernest*. New York: Scholastic.

Rockwell, A. (1993). *Pots and pans*. New York: Macmillan.

Shaw, N. (1992). *Sheep out to eat*. Boston, MA: Houghton Mifflin.

Extenders

Music/Cognitive: Try this variation of the lemon song:
 "Where do you hide a sour lemon, ·
 where do you hide it, where do you hide it,
 where do you hide a sour lemon,
 where do you hide that fruit?"

Ask the children to sit in a circle. Have them take turns hiding a real lemon behind their backs while everyone sings the song. Have the children take turns sitting in the middle of the circle with their eyes covered, while the song is sung and the lemon is hidden. Can they guess where it is? If you like, and if your students feel comfortable with this, have the child who hides the lemon sing the song alone and stop as soon as the fruit is hidden, so that the child in the middle must use the sense of hearing to guess who has it.

Sensory/Science: When you conduct the shaking experiment with tin cans, vary it by putting different amounts of different materials in the tins: (e.g., marbles, beads, straw pieces, paper clips, stones), and have the children experiment with guessing whether the items are heavy and light. At first they may guess only by listening to tins being shaken, and then by listening *and* shaking.

Language/Art: Provide construction paper, scissors, magazine pictures, a hole puncher, and glue, and have the children make their own Tricky Pictures.

Field Trip: Visit a restaurant. Before you go, talk about how you might use each sense once you are there.

TATER TIME

Attention Getter: When the children are gathered, ask them to close their eyes, open their mouths, and stick their tongues out. Tell them you are going to put something on their tongues that is good to eat and that it will give them a clue about what you are all going to talk about during the week. Put a small potato chip on each of the children's tongues. If or when your students guess that they are eating potato chips, ask them what chips are made of, and then ask them to guess what you will be working with for the next few days.

Several activities in this unit utilize potatoes as a material for art or sensory exploration. Because of world hunger, some educators are reluctant to use food for play in early childhood curricula. Other educators feel the experiences with food are very worthwhile. In any case, I have included the activities so that you can choose what is right for you.

A Potato Is Stored Food

Science

WHY we are doing this experiment: to enable children to observe that a potato is stored food that can nourish a growing potato plant.

WHAT we will need:
  Potatoes
  Magnifying glasses

HOW we will do it: Buy some potatoes about a month before you plan on doing this unit, and store them until protuberances begin to grow from the potato eyes. Bring these in, along with a few potatoes with eyes but no growths, and set them on an activity table with magnifying glasses for the children to examine.

During an Attention Getter time, show the children the potato eyes, and then one of the potatoes with a shoot growing from an eye. Emphasize how important it is for the children to be very careful and gentle so that the root does not break off. Pass the potato around the group for the children to hold and examine. Ask: "How can this root be growing when the potato isn't planted in the ground?" Encourage the children to express their ideas. (The potato itself is stored food, and the growing root is fed by the potato until it can find water in the earth.) Show the children the growing potatoes and magnifying glasses on the activity table, and invite them to examine them. Before the children approach the table, talk together about what will happen if the potatoes are handled roughly. (The growing roots will break off.) If you think your students will have trouble with this, make a reminder sign or put the potatoes in or under clear containers.

Growing Potatoes

Science

WHY we are doing this experiment: to enable children to see the initial stages of growth of a new potato plant and to observe how a new plant begins to grow from the original potato.

The facts of the matter: The long white hairs that will grow are roots and the rhizome. The roots extend downward and absorb nutrients. The rhizome swells when the new plant finishes growing and becomes the tuber, or actual new potato. From the same potato eye from which the roots grow, a green shoot will grow upward. This will become the plant stalk from which leaves will eventually emerge.

WHAT we will need:
  Potatoes (one per child)
  Plastic containers (empty peanut butter
    containers are ideal)
  Toothpicks
  Small pitchers of water
  Masking tape
  Marker

HOW we will do it: You will need a place to leave the potato experiments for at least two weeks. Put the toothpicks, plastic containers, potatoes, and pitchers of water on the table. Tear off masking tape strips for name tags and stick them onto the edge of the activity table. Using the marker, draw a line around each potato at about the one-third mark so the children know where to insert the toothpicks.

Ahead of time, prepare a sample potato-growing experiment. Stick three toothpicks into a potato at equal intervals around the line. Rest the potato on the rim of a container so that approximately one-third of the potato will be in water when the container is filled.

During an Attention Getter time, show the children your experiment. Say: "Remember the roots we saw growing from the potatoes? I was wondering how big those roots would get if they had some water, and whether or not a new potato plant would grow from this potato. So I put toothpicks in the potato right where the line is, so that part of my potato could sit in the water." Show the children the materials on the table, and invite them to prepare their own potato-growing experiments. Ask them to predict what the potato will do. Encourage them to write their names on the masking strips and to label their potatoes.

Potato Observation Books
Science/Language

WHY we are doing this experiment: to enable children to make a daily record of the results of the previous experiment; to develop all components of language arts: reading, writing, speaking, and listening.

WHAT we will need:
Writing sheets (provided below;
    photocopy and enlarge
    for your use)

Blank paper
Light brown construction paper
Markers
Crayons
Pens
Preparation:
    Scissors
    Stapler

HOW we will do it: To prepare, make several copies of the potato observation sheets provided. Make books out of these, using a blank piece of brown construction paper for the covers. Make one for each child. Also, make some books of blank paper for children who want to create their own observation records. Put the books and writing/drawing utensils on an activity table near the potatoes.

Make a sample observation book. Write a title on the cover and draw pictures on it. Draw and write your observations about your potato in the first space. During an Attention Getter time, show the children your observation book. Together, read/interpret the words on the

Science Experiment: Potato Roots

Write the appropriate number next to each day, and make as many photocopies as you need. Children can draw, write, or dictate observations.

Day
I see:

Day
I see:

Day
I see:

Day
I see:

cover and inside the book. Discuss the drawings. Show the children the materials on the table and invite them to start their own observation books, either with the prepared writing sheets or with blank paper. Leave the books out for the duration of the experiment so that children can work on them as and when they choose. After a week or two, have a show-and-tell of observation books by having a few children talk about their books every day.

Growing New Potatoes
Science

WHY we are doing this experiment: to enable children to observe that new potatoes grow from a potato.

WHAT we will need:
    Potatoes with eyes
    Gardening tools
    Plot of earth
    Rulers
    Camera
Preparation:
    Sharp knife

HOW we will do it: You will not see the fruits of this experiment until six to ten weeks after you plant the potato pieces. Even so, it gives children the opportunity to compare the grown potato plant, complete with new potatoes, to the small piece originally planted. If necessary, loosen the dirt of your plot before the children plant their potato pieces. Use a knife to cut pieces of potato, making sure each piece has one or several eyes. During an Attention Getter time, ask the children if they remember what grows out of the potato eyes. Talk again about what you discovered in the Growing Potatoes and A Potato Is Stored Food experiments. Ask the children to predict what will grow from the potato eyes.

Show the children the plot of earth and potato pieces and ask them if they have any ideas about what they could do with them.

Before planting the pieces take photographs of them so that you can compare them to the grown plants, in later weeks. Invite your students to use the gardening tools to plant the potato chunks. They should be planted 12 cm to 18 cm (5″ to 7″) deep in the ground. When the stalk is about 50 cm tall (20″), the plant has probably grown tiny new potatoes. This can take six to ten weeks.

Check on your plants once a week, and if there has been no rain, water them. When your stalks are the right height, dig one plant up. If you see new potatoes, you can dig up the rest of the plants; otherwise, let them grow for a few more weeks. Shake your plants well to dislodge clods of earth and insects. Ask the children to compare a whole potato plant with the photographs of the original potato pieces. What are all the different parts which grew out of the tiny eyes on the original potato chunks? Invite the children to touch the plant stalk, leaves, roots, and new potatoes.

Potato Discovery Center
Science

WHY we are doing this activity: to enable children to examine, at the same time, a number of potatoes at different stages of growth in the cycle of potato reproduction.

WHAT we will need:
    Potatoes
    Magnifying glasses
    Index cards
    Markers

HOW we will do it: About six to ten weeks ahead of time, plant a few potato pieces with eyes. Three to five weeks ahead, plant a few more. About one week or one and one-half weeks ahead, put a few potatoes in jars of water, as described in the Growing Potatoes activity.

When you are ready to set up the discovery center, dig up the potato plants, making sure you dislodge all clods of earth, and shake the plants well so that insects are left behind. Put these potatoes, the potatoes in jars, and the regular, unsprouted potatoes in your discovery center, along with the magnifying glasses. Use the index cards and markers to label how many weeks each plant grew before it was dug up. Invite the children to examine and compare the potatoes and to discuss their observations.

Potato Badges

Language

WHY we are doing this activity: to develop speaking and listening skills.

WHAT we will need:

Light brown construction paper

Potato shapes (provided below; photocopy and enlarge for your use)
Double-sided tape or masking tape loops (sticky side out)
Markers

HOW we will do it: Use the potato shape provided to cut out potato badges from the construction paper. Leave some badges blank, and on the rest, write: "Ask me about my potato science experiment!" Use double-sided tape or a masking tape loop to stick one onto yourself. Put the rest of the badges and the markers on an activity table. Stick the tape strips around the edge of the table. If the children do not notice your badge, read it to them and show them the other badges on the table. Encourage children to write their own words on the blank badges if they wish. If necessary, help your students secure their badges to their clothes.

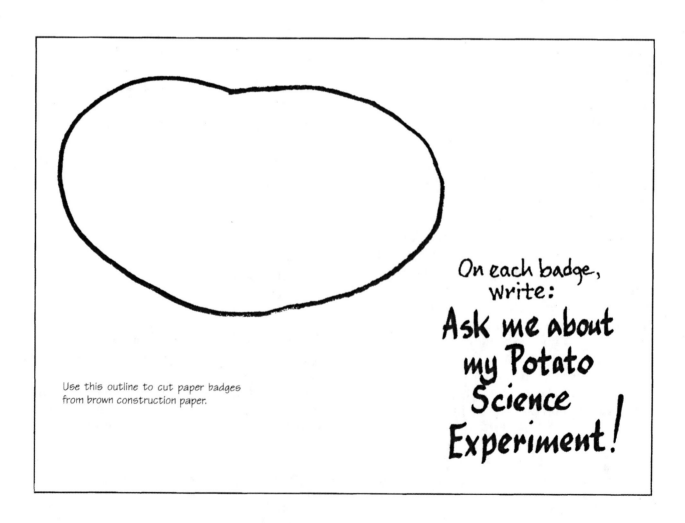

Use this outline to cut paper badges from brown construction paper.

On each badge, write:
Ask me about my Potato Science Experiment!

150

What Is in a Potato?

Science

WHY we are doing this experiment: to enable children to observe that there is sugar (starch) in potatoes.

The facts of the matter: A starch molecule consists of long chains of sugar molecules.

WHAT we will need:
Iodine
Potato slices (raw)
Medicine droppers
Small containers
Potato peeler

HOW we will do it: Put all materials out on an activity table. During an Attention Getter time, explain to the children that the food which is stored in potatoes, and feeds a growing plant, is a type of sugar, also called a *starch*. Show the children the iodine and ask them to notice what color it is. Explain that the children are going to prove that there is starch in potatoes by using iodine. Iodine will turn to a blue-black color when it is put on starch. It is very important that you say: "Iodine is very, very dangerous to taste or drink. We're only going to drop it on our potato slices. We're not going to put the iodine, or the potatoes with iodine, anywhere near our mouths, because that would be dangerous." Discuss this. (The best way to supervise this activity is to sit at the table at where the experiment takes place, so that you can watch the children for the duration of the experiment.)

Invite the children to conduct the experiment. Discuss their results. What happens when they put iodine on the potato pieces? What does this tell us about what is in potatoes? Use a potato peeler to carve off large pieces of peel onto which the children can drop iodine. Do the children think a potato peel has sugar in it?

Potato Foods (Part 1)

Science/Field Trip

WHY we are doing this activity: to enable children to observe how many different ways there are of using potatoes for food.

WHAT we will need:
Supermarket
Money

HOW we will do it: You may want to contact the manager of the supermarket you will be visiting with the children and discuss your plans, especially if you have a large group.

Before you leave your home or classroom, let the children know that they are going to visit a supermarket to see how many foods they can find in the store that are made out of potatoes. Ask the children how many foods they can think of that you can buy in a store and which are made from potatoes. When you actually enter the supermarket, remind your students to look for bags or boxes or packages of food that might have potatoes in them.

Go to the produce section first and let each child pick out a small potato to carry as you tour the store. This helps the children make the connection between a fresh potato and the processed foods they will find, which are made out of frozen, dehydrated, fried, or otherwise processed potatoes. Here are some potato foods you might spot and discuss: boxes of potato casseroles (e.g., scalloped, cheddar), potato flakes or powdered potatoes for instant mashed potatoes, frozen hash brown potatoes, frozen french fries, frozen Tater Tots, potato chips, Pringles, and potato sticks. Buy a few of these to take back to your home or classroom. Include dehydrated potato casserole packages, as well as flakes and powder for instant mashed potatoes, so that the children can experiment with what happens to them when water is added. (See the Dried Potatoes activity.)

Potato Foods (Part 2)
Science/Sensory

WHY we are doing this project: to enable children to observe the difference between processed potatoes and fresh potatoes; to discuss relative nutritive value; to provide a taste experience.

WHAT we will need:
Potato food samples
Fresh potatoes
Potato peeler

HOW we will do it: Ahead of time, boil, bake, or roast fresh potato pieces with their skins. Make enough for each child to sample them. Prepare the processed potato foods that require cooking and put these and the other processed samples from the store on a small table for tasting. Put some uncooked samples from each package or box next to the corresponding cooked sample so the children can see how the potatoes changed. If you work with three- and young four-year-olds, you may want to put the uncooked samples in clear plastic cups and tape them shut with packaging tape so that they are not eaten, or discuss with the children which samples can be eaten and which cannot.

During an Attention Getter time, show the children the potato samples available for tasting and the original packages from which each one came. Discuss how many samples each student can have. What will happen if someone takes more? (There will not be enough for everyone.) Hold up a fresh potato and ask the children: "How is this potato different from these potato foods? After this potato was dug out of the earth and washed off, do you think anything else was done to it? What about these potato foods we bought in the store? Are they the same as this potato? How are they different?" Talk about the fact that a vegetable that has not been cooked or frozen or dried is called *fresh*. Vegetables that have been cooked or dried or frozen are called *processed*. Pass the fresh potato around the group while you repeat the word *fresh*. Pass a dried potato slice from a casserole package around the group while you say the word *processed*. Ask the children

whether they think fresh or processed foods are better. Discuss. Which of these is better for helping your body grow: a fresh potato, or the potato snack junk foods? Discuss.

Using the potato peeler, carve off a potato skin peel for each child to hold. Tell the children that there are a lot of vitamins in potato skins. Ask them if they can think of ways of eating potatoes that include the skins (e.g., baked potatoes, hash browns with the skin on, mashed potatoes with the skin included). Look at your store-bought potato food samples. Does it look like there is any potato skin in any of them? What did people have to do to the potatoes to make each of the processed potato foods you bought at the store? (For example, take the water out; make the potatoes into flour or dried chips or flakes; peel, slice, fry, and salt; peel, dice and freeze.) Tell the children the simple things you did to cook your fresh potato pieces, and the fact that the less you do to food, the better it usually is for your body. When the children try the different samples, ask them if your cooked potato pieces feel different in their mouths than the more processed potato foods. Introduce the word *texture*. As the children try all of the different kinds of potatoes, encourage them to express their observations.

How Much Water Is in Potatoes?
Science

WHY we are doing this project: to enable children to discover that potatoes have a high water content.

WHAT we will need:
Freestanding graters
Small- to medium-sized potatoes
(or large potato chunks)
Carrots

HOW we will do it: Set out the potatoes and graters on an activity table. Ask the children if they think potatoes have any water in them. Invite the children to grate the potatoes and to squish between their fingers the potato pulp produced by the grater. How does it feel? What does this tell us about what is in potatoes? After the children have squeezed and squished the pulp, put carrots on the table. When carrots are grated, what happens? Invite the children to compare how the potato pulp feels with how the carrot pulp feels. What does this tell us about which vegetable has more water in it?

Tip: Save some of the potato pulp for the following activity.

Dried Potatoes and Water

Science

WHY we are doing this project: to enable children to add water to the various dried potatoes purchased in the Potato Foods activity; to enable them to compare the dried and hydrated textures of the potato foods.

WHAT we will need:
Boxes of potato casserole mix (the kind with dried potato slices)
Potato flakes
Potato powder (for instant mashed potatoes)
Medicine droppers
Containers of warm water
Containers of cold water
Containers for mixing potatoes and water

HOW we will do it: Put all the materials on the activity table. During an Attention Getter time, take out the potato pulp from the previous project and encourage the children to touch it again. How does it feel? What does it tell us about what is in potatoes?

Show the children the dried potato foods on the table and ask them how they are different from the pulp. Invite them to mix the dried mashed potato powders with water. Potato slices from casserole packages can be soaked overnight. The next morning, compare the soaking slices to the dried ones. What happened? Invite the children to compare the hydrated potato foods to the fresh potato pulp, and encourage tactile exploration of all materials.

What Grows from a Sweet Potato?

Science/Crafts/Construction

WHY we are doing this project: to promote self-esteem by enabling children to prepare their own materials for a science experiment; to promote cooperation between students by facilitating a group project; to provide fine motor activity; to enable children to observe what happens to a sweet potato as the vine flourishes; to provide recording and measuring opportunities.

WHAT we will need:
Two wooden meter sticks for each trellis
Popsicle sticks
Wood glue (e.g., Elmer's)
Glue brushes
Butcher paper
Paper
Markers
Pens
Rulers
Fresh sweet potato (not kiln-dried)
Plastic jar
Toothpicks
Water

HOW we will do it: The day before this activity, glue two or three Popsicle sticks across both meter sticks like the rungs of a ladder. The Popsicle sticks should be at about 5 cm or 6 cm (2" or 3") intervals. This will form the basis for a trellis. Invite the children to continue gluing Popsicle sticks onto the meter sticks.

The trellis that the children complete will not be as even as a ladder, and in fact, may not even be the shape of a ladder, but having a few Popsicle sticks already glued on will suggest what is needed.

After the glue has dried, ask the children where the trellis could be propped. Explain that you are going to put a sweet potato at the bottom of the trellis and it will need light. Together, prepare a place for the experiment. If possible, place your sweet potato and trellis on the floor or ground. This will make it easier for children to mark the growth of the vine on the chart next to the trellis. Make this chart out of butcher paper, by cutting out a strip which is about 20 cm (8") wide, and as long as the trellis. Post the chart directly next to the trellis. (You may decide to set up several trellises and sweet potatoes, or you may decide to set up only one.) Draw a line around your sweet potato about one-third of the way above the tapered end.

During an Attention Getter time, encourage the children to put toothpicks in the line around the sweet potato, to pour water into the container, and to prop the sweet potato in the container with the tapered end downward, so that one-third of it is covered in water. Ask the children where they should put the sweet potato so that if a vine grows, it will climb the trellis.

A vine may grow for months before the sweet potato collapses in on itself. When the buds and roots begin to appear, move a small table near the chart. Set out the rulers, markers, pens, and paper. Invite the children to mark off, on the chart, the height of the vine as it climbs the trellis, and to measure the height with the rulers. Children can also draw pictures of the sweet potato and vine and record descriptions by using the paper, markers, and pens. How does the vine use the trellis? Encourage the children to examine the tendrils closely to see how they wrap around the sticks. What happens to the sweet potato eventually? When it collapses, you can plant the vine outdoors if you wish.

Potato People
Art/Crafts/Fine Motor

WHY we are doing this project: to facilitate creative expression and to develop fine motor skills.

WHAT we will need:
> Potatoes (various shapes and sizes)
> Toothpicks
> Olive slices (rings)
> Carrot pieces (tips and stumps)
> Parsley bunches
> Celery pieces, some with leaves
> Cherry tomatoes, halved

HOW we will do it: Put all materials out on the activity table. Make a sample potato person or animal by pinning vegetable pieces onto a potato with toothpicks. Parsley makes good hair, olive slices and halved cherry tomatoes make good eyes, carrot stumps can be used as legs, and tips can be used as noses. Be creative!

During an Attention Getter time, show the children your potato person or animal, and show them the materials on the activity table. Encourage them to make their own potato creations.

The Spud Fell Out of the Barrel
Music/Movement

WHY we are doing this activity: to develop cognition by memorizing words and connected actions; to develop an appreciation and enjoyment of singing together.

WHAT we will need:
> Song: "THE SPUD FELL OUT OF THE BARREL" (To the tune of "The Bear Went Over the Mountain")

> "THE SPUD FELL OUT OF
> THE BARREL"
> The spud fell out of the barrel, the spud fell out of the barrel, the spud fell out of the barrel, and rolled off far away.
> It tumbled down a mountain . . .
> and hit up against a rock.
> It flew into a window . . .
> and under a kitchen knife.
> It got chopped up in pieces . . .
> and went into the soup.
> The soup went in my tummy . . .
> and was so very good.

HOW we will do it: Explain to the children that *spud* is another name for a potato. Sing the song with the following motions: For "The spud fell out of the barrel," make one arm into a barrel opening and the fist of the other hand into a potato. Have your spud fall out of the barrel, and when you come to "and rolled off far away," make your potato fist turn over and over.

For "It tumbled down a mountain," turn both of your hands over and over each other. For "and hit up against a rock," slap one palm against another. For "It flew into a window," arch your arm to make a window and make your fist fly through it. For "and under a kitchen knife," make chopping motions with the side of one hand and make your other fist fall under it.

For "It got chopped up in pieces," make both hands into knives and pretend to chop. For "and went into the soup," pretend to scoop up the potato pieces and put them into a pot. For "The soup went in my tummy," pretend to eat soup from a bowl with a spoon. For "and was so very good," rub your stomach.

If you like, get a big soup pot to put in the middle of your circle, and use this to throw your imaginary potato pieces into and to eat your imaginary soup from.

Making Mashed Spuds
Science/Math/Cooking/Small Group Project

WHY we are doing this activity: to allow children to observe how hot water changes the texture of potatoes; to provide experience with measuring and counting; to develop self-esteem by enabling children to eat a snack they have prepared themselves; to develop cognition by recalling ingredients and sequence of preparation.

WHAT we will need:
   One potato (half-peeled)
   Several potatoes (peeled and chopped
      into roughly 2.5 cm (1") square pieces
      [for faster cooking])
   Recipe sign cards (provided below;
      photocopy and enlarge for your use)
   Plastic soup bowls
   One tablespoon

One large serving spoon
Forks
Milk
Softened butter (cut into pats)
Salt
Masking tape
Colander

HOW we will do it: On a table, set out the half-peeled potato for the children to examine. Also set out a large bowl for the cooked potatoes, a bowl of milk, a bowl of butter squares, and a bowl of salt. Place the appropriate recipe card behind each one.

Have the uncooked, peeled, and chopped potatoes ready in another bowl. Boil or microwave a pot of water. In order to prevent children from having access to the stove while the water is boiling, put a masking tape line on the floor to indicate where the children may not cross, or block off the area with chairs.

Have the children wash their hands first. Take out a few of the peeled potato chunks and encourage the children to touch them. Leave them out on the table. Let your students know you are going to put the potato pieces into the boiling water. Ask them to predict whether the potato pieces will change in the hot water and if so, how. Before you put the potato pieces in the pot, ask the children to put their hands behind their backs. Scoop out a mug of the hot water. Ask the children if they can see the steam from the water. Why is there .steam? Put the potato pieces into the boiling water. When the potato pieces are tender, pour the contents of the pot into a colander over a sink, and then empty the pieces into a large bowl. Take a few chunks out and run them under cool water. Encourage the children to compare the few uncooked chunks with the cooked ones. Leave both kinds out for the duration of free play activities. How did the heat of the boiling water change the potato?

When the rest of the potato chunks have cooled off enough for the children to safely handle them, put them in a bowl with the serving spoon and then place them on the table with the other ingredients. Put the appropriate sign next to the potatoes. Encourage each child to take a plastic soup bowl, go around the table, and take the amount of each ingredient that

each sign indicates. Invite them to sit down with a fork, and mash all their ingredients together, and to then eat what they have made. As you eat your snack together, talk about the sequence of events in making your mashed potatoes, to reinforce this for the following language activity. Save the recipe cards you used for this same purpose.

Class Book: Recipe for Mashed Spuds
Language/Small Group Project

WHY we are doing this project: to facilitate group cooperation and interaction; to develop all components of language arts: reading, writing, speaking, and listening; to develop self-esteem through use of a one-person work table.

WHAT we will need:
   Light brown construction paper
      (large sheets)
   Markers
   Crayons
   "One person may be here" sign
      (provided on page 165; enlarge and
      photocopy for your use)
Preparation:
   Stapler
   Scissors

HOW we will do it: To prepare, put four or five sheets of construction paper on top of each other, and staple down one side. Cut a potato shape out of the paper, making sure that you do not cut the staples off of the stapled side (in other words, make sure that even after you cut your potato shape, the pages still hold together).

Ask the children if they know what a *recipe* is. Hold up the recipe cards you used in the mashed potato activity. Encourage the children to tell you how they made the mashed potatoes. Say: "A *recipe* tells people how to make something they can eat. I thought we could put our class recipe for mashed potatoes in this book." Take dictation from the children about how they made their mashed potatoes and write their words in the book. When they are finished, read the recipe back to them. Put the book, crayons, and markers at a one-person work table. Discuss the "One person may be here" sign, and let the children know they can take turns illustrating the book. When the book is finished, put it with the rest of your books so that the children may look at and read it.

Mashed Mush
Sensory

WHY we are doing this activity: to facilitate a sensory experience with an unusual play material.

WHAT we will need:
>Prepared instant mashed potato mix
>Tubs or sensory table
>Food coloring (optional)
>Containers
>Spatulas
>Spoons

HOW we will do it: Put the mashed potato mix and the play materials into the tub or sensory table. Add food coloring if you like, and encourage the children to explore.

Tip: Save some of the uncolored mashed potatoes for the Mashed Potato Art activity.

Potato Count
Math

WHY we are doing this activity: to practice rational counting; to familiarize young children with the concepts of *large, larger, largest*; to develop self-esteem and a sense of autonomy through use of a one-person work station.

WHAT we will need:
>Potatoes (large, medium, and small)
>Three containers of graduated sizes (e.g., a plastic 900 g (2 lb.) yogurt container, an 1800 g (4 lb.) yogurt container, and a basin)
>"One person may be here" sign (provided on page 165; enlarge and photocopy for your use)
>Writing sheets (format provided on page 158)
>Blank paper
>Pens
>Activity sign

HOW we will do it: Fill the containers with potatoes. If you work with younger children who are not counting past ten, adjust the size of the containers and the number of potatoes accordingly. Make writing sheets based on the provided format, but draw symbols that represent the sizes and shapes of the containers you actually use. Set out several copies of these, along with the pens and blank sheets of paper, next to the potatoes. Make an activity sign that says: "How many potatoes?" Draw some potatoes above the word. Post this and a "One person may be here" sign near the activity table.

During an Attention Getter time, read/ interpret the activity sign and writing sheets together. Show your students the blank paper that can also be used to record the potatoes counted. Discuss the "One person may be here" sign. Encourage the children to explore the materials.

How many potatoes?

I count _____ potatoes in the

large container.

I count _____ potatoes in the

larger container.

I count _____ potatoes in the

largest container.

Mashed Potato Art

Art

WHY we are doing this activity: to facilitate creative exploration with color; to develop fine motor skills.

WHAT we will need:
Instant mashed potatoes
Medicine droppers
Small containers
Water
Food coloring
Basin
Bowl of water

HOW we will do it: Cook the mashed potatoes. Put a small amount in each container, and put the containers in front of the chairs at the activity table. Using food coloring, mix up red, blue, and yellow water, and place all three colors around the table so that the children will have access to them.

Encourage the children to use the medicine droppers and colored water to color their mashed potatoes. What happens when they drop one color on top of another? Tell the children they may dump the colored potatoes into the basin when they want to start a new batch, and that they may use the bowl of water to rinse out clogged droppers.

Potato Prints
Art

WHY we are doing this activity: to facilitate creative expression.
WHAT we will need:
    Potatoes (large and small)
    Paint
    String
    Rubber bands
    Butcher paper
    Shallow pans (for the paint)
    Newspaper
    Paper towels
Preparation:
    Knife

HOW we will do it: Spread many layers of newspaper on an activity surface and spread the butcher paper sheets out on top. Set out shallow pans of paint, so that all children will have easy access to them. The paint should not be too watery or the potatoes will not make good prints. Cut the potatoes in half, vertically and horizontally, and sandwich them between paper towels for a few minutes so that water is absorbed; this produces better prints. To vary the prints the potatoes make, wrap rubber bands around some, string around some, and use the knife to cut grooves in others.

Provide a good variety of different shapes and sizes. Put the potatoes in the pans with the paint, and encourage the children to make potato prints.

Potato Pancakes
Multicultural/Small Group Activity

WHY we are doing this activity: to introduce children to a Jewish tradition.

WHAT we will need:
    Book: Hirsh, Marilyn, *Potato Pancakes All
        Around: A Hanukkah Tale*

6 medium potatoes, peeled and grated
2 tablespoons flour
2 tablespoons heavy or light cream
2 eggs, beaten
Salt to taste
8 tablespoons bacon fat or oil
One raw, unpeeled potato
Paper towels
Electric frying pan
Chairs or masking tape
Small paper plates
Plastic forks
Napkins

HOW we will do it: Peel and grate the potatoes well ahead of time. Keep the grated potatoes in water to prevent them from turning brown, but be sure you squeeze the water out thoroughly before cooking. You can do this by wrapping the grated potato in strong, absorbent paper towels and squeezing and twisting all the excess water out into the towels. Put all the other cooking ingredients on the activity table. Plug in your electric frying pan, and make a barricade of chairs around it. Depending on your students, you may choose to use a masking tape line that they may not cross.

During an Attention Getter time, read *Potato Pancakes All Around*. There is no large, clear illustration of a potato in this book, so I like to let the children pass a potato around while I read the story. Show the children the cooking ingredients on the table and ask them what they could do with them. Have the children wash their hands and discuss the importance of not sneezing into the food, or putting fingers into mouths, noses, and ears while preparing the pancakes. Show the children the frying pan and the masking tape line or chair barricade. Let them know they may not cross the line, and say: "We're going to cook our potato pancakes in this frying pan and it's going to get very hot. What would happen if you touched it?"

Heat the frying pan. Put the grated potatoes in a bowl and add the flour, cream, egg, and salt. Let the children take turns stirring the mixture. Put about four tablespoons of the mixture in the pan to make each pancake. Cook both sides until browned. Let the pancakes cool a little, then enjoy!

Tater Toss

Gross Motor

WHY we are doing this activity: to develop hand-eye coordination; to develop the large muscle group.

WHAT we will need:
    Large tubs or baskets
    Small potatoes
    Masking tape

HOW we will do it: Set the tubs or baskets at various places around the room. Put a masking tape line down several feet away from each one. The distance between the containers and the tape should depend on the age of your children—further away for older students, closer for younger children. Put several potatoes by each masking tape line. Encourage the children to toss the "taters" into the containers, and sincerely praise all their efforts. Before the activity, you may want to talk about what would happen if a potato were thrown at someone, and how important it is not to do this.

Greengrocer

Dramatic Play/Math/Language/Social Studies

WHY we are doing this activity: to facilitate an understanding of what a greengrocer does; to practice rational counting with pennies and nickels; to develop cognition by distinguishing between fruits and vegetables; to facilitate weighing and comparing; to develop all components of language arts: reading, writing, speaking, and listening.

WHAT we will need:
    Balancing scales
    Toy cash registers
    Pennies and nickels
    Potatoes
    Other fruits/vegetables: (e.g., carrots, onions, apples)
    Produce advertisements from newspaper supplements
    Blank "Special" or "Sale" cards (ask your local supermarket to donate some)
    Checkbooks and receipt forms (pattern provided on page 161; photocopy and enlarge for your use)
    Pens
    Markers
    Baskets or bags
    Children's aprons

HOW we will do it: Set up all materials to simulate a greengrocer's stand. Make several copies of the check patterns provided and cover them with construction paper to simulate checkbook covers. Also make copies of the receipt forms and staple them into a notebook. Post the produce advertisements on the walls around the store and set out blank "special" and "sale" signs with markers.

During an Attention Getter time, ask the children whether a potato is a fruit or a vegetable. Hold up one of each of the fruits or vegetables you will be using in the play greengrocer's, or use photographs, and ask the same question. If there is confusion, tell the children that, generally, fruits are sweet to eat and, generally, vegetables are not. Ask the children if any of them have ever been to a store that sells only fruits and vegetables. Tell them that a person who runs a store like that is called a *greengrocer*. Invite the children to explore the play greengrocery.

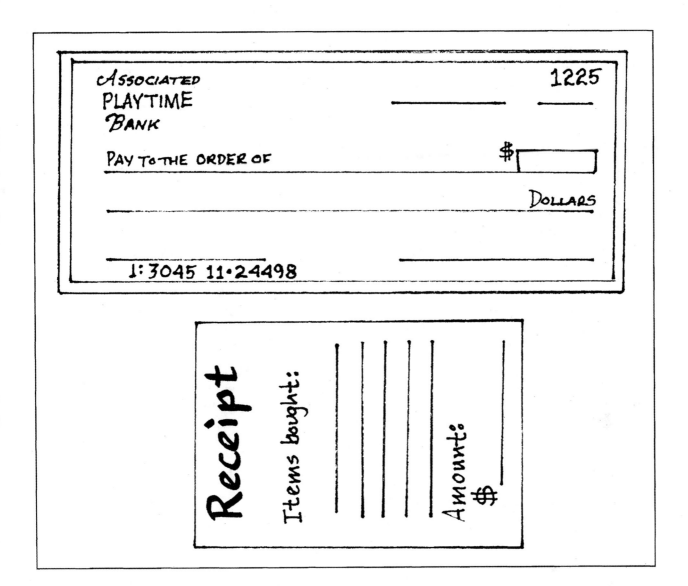

Literature

Symbol Key: *Multicultural
              +Minimal diversity
              No symbol: no diversity or no people

Brenner, B. (1992). *Group soup*. New York: Viking.

Hirsh, M. (1978). *Potato pancakes all around: A Hanukkah tale*. Rockaway Beach, NY: Bonim Books, Hebrew Publishing.*

Lobel, A. (1967). *Potatoes, potatoes*. New York: Harper & Row. (This excellent book addresses peace and anti-war issues.)

Selsam, M. E. (1972). *More potatoes!* New York: Harper & Row.

Wake, S. (1990). *Vegetables*. Minneapolis, MN: Carolrhoda Books.*

Watts, B. (1987). *Potato*. Morristown, NJ: Silver Burdett Press. (Excellent picture book with good photographs.)

Extenders

Social Studies: If possible, take a trip with the children to the local greengrocer's. Ask the greengrocer to give you a tour of the store so that you can see where fruits and vegetables are sorted and washed. What does the produce arrive in? How many different kinds of potatoes does the store sell? If you do not have a convenient greengrocer's to visit, tour the produce section of your local supermarket, and ask the manager to show you where the produce is delivered, washed, and sorted in the back of the store.

Math: Gather a diverse collection of potatoes: red, white, russet, baking potatoes, new potatoes (if in season), sweet potatoes, and big and small potatoes. Encourage the children to sort them. For younger children, get them interested by telling the sorting story in the Woodworks unit, but substitute the word "potato" for "nail."

Language: When you explore the store-bought processed potato foods, make a language chart of all the potato dishes the children eat at home.

Gross Motor: Play "Hot Potato." Have the children stand a few feet away from each other in a circle, and toss a potato to each other while music is being played. Ask them to pretend that the potato is very hot and they will burn their hands if they hold it for more than a few seconds. Turn the music off suddenly. Who is holding the hot potato? Let that child turn the music off the next time.

COLOR CAPERS

Attention Getter: Gather the children together before they begin exploring the activities, and have a large, clear jar of water available. If you wish, have some relaxing dream-like music playing in the background. Using food coloring and a medicine dropper, let drops of different colors fall into the jar. With the children, watch the colors diffuse while they curl and bend, making streamers in the water.

Primary Patterns

Science/Cognitive

WHY we are doing this project: to reinforce the concept of primary colors; to help children understand that red, yellow, and blue make other colors; to develop fine motor skills.

Developmental differences: Three- and young four-year-olds often do not remember the term *primary* or which colors are categorized this way, but their hands-on, child-initiated exploration of this activity will pave the way for later understanding. If you want to reinforce the concept with kindergartners, print a chart of the primary colors, with a patch of the appropriate color next to each color's name, and hang it up in your room.

WHAT we will need:
>   Clear contact paper
>   White cardboard or construction paper
>   Medicine droppers
>   Food coloring
>   Muffin pans or other sectioned containers

HOW we will do it: To prepare, cut the paper in squares about 25 cm x 25 cm (5" x 5") and cover both sides with contact paper. An easy way to do this is to cut strips of contact which are about 14 cm x 28 cm (5½" x 11"), place the paper or board on one end, fold the rest of the contact over it, and seal the edges to waterproof the paper or cardboard inside.

Using medicine droppers, the children drip red, yellow, and blue food coloring on the boards. Find out what colors are produced when these original colors are mixed. If you do not have time to make the boards, children can experiment in small containers; however, they really enjoy using the boards because individual drops on plastic contact stay self-contained and retain very distinct shapes. In your muffin tins, put small amounts of water separately colored with red, yellow, and blue food coloring, and place them on the table so that all children can reach them. As you watch the children experiment, talk about the fact that they are using primary colors, and comment on the new colors they are creating. Use the words "science experiment" so that the children know *this* is science!

The Primary Song

Music/Movement/Science/Cognitive

WHY we are singing this song: to reinforce the concept of primary colors musically; to help children feel good about their singing voices; to provide a cognitive exercise in identifying red, yellow, and blue; to facilitate a memory game.

WHAT we will need:
>   Construction paper (red, yellow, and blue)
>   Contact paper (clear)
>   Song: "THE PRIMARY SONG" (to the tune of "Twinkle Twinkle Little Star")

"THE PRIMARY SONG:"
Primary, primary what are you?
I am red or yellow or blue.
Primary, what can I do with you?
Mix us into colors that are new.
Primary, primary what are you?
I am red or yellow or blue.

HOW we will do it: First cut red, blue, and yellow shapes out of construction paper and cover with contact paper. Cut enough shapes for each child to have one of each color. Next sing this song together, singing the third and sixth lines slowly so that you have time to hold up the appropriate paper shape when that color is mentioned.

Sometimes, as time goes by and a song becomes very familiar to children, it is fun to alter the routine a little, by using variations. Here is one option: decide together on one color to leave out when you sing the song. Can everyone remember not to sing that word? You can vary the game even more by deciding to hold up the color but not sing it, and to increase the number of colors you decide to omit. Have the children make suggestions about other variations of this game, and let them take turns holding up the color shapes as you sing.

Acetate Action

Science/Cognitive

WHY we are doing this project: to reinforce the concept of secondary colors through the use of an interesting material; to develop a sense of autonomy through use of a one-person work station; to develop creativity by providing children with the opportunity to make their own colors.

WHAT we will need:
- Acetate papers including red, yellow, and blue
- or: transparent plastic folders (same colors as above)
- "One person may be here" sign (provided below; photocopy and enlarge for your use)
- Secondary color sign (directions to follow)

HOW we will do it: Acetate papers are available at graphic art supply stores. If you mention that the papers are for young children, businesses may donate supplies. If you work for a school which sends a newsletter to parents, you can sometimes offer free advertising in your publication in return for materials.

To prepare, cut the folders or acetates into single sheets, and lay them on the one-person table. Print a language chart that says: "Orange is a secondary color. Green is a secondary color. Purple is a secondary color." (Language charts should clarify, not confuse, so depending on how firmly you feel your children have grasped the concepts so far, you may wish to create a chart in the format of color equations: "Red and yellow make orange" and so forth. Or you could print a chart with this format: "Primary red and primary yellow make secondary orange." Decide which one you think will be most helpful to your students.) For all charts, put a patch of the corresponding color next to each color's name.

Hang the sign up on the wall in front of the activity table, and also put up the "One person may be here" sign. Help the children interpret/ read these signs when they have gathered together as a group, before activity exploration, and discuss the sign's meanings. (For example, ask the children: "Does that mean that two people can be there? Does that mean that three people can be there? What can you do if you want to be there and someone else is already there?") Then take acetate sheets of two primary colors and ask the children to predict what they will see when you lay one over the other. After the children have made their predictions, invite a student to lay the two sheets over each other. What happens? Tell the children that what they see is a secondary color. Point to the color chart and ask them which secondary color the two sheets created. Put only red, blue, and yellow sheets on the activity table for the first few days. As the children experiment, comment on the secondary colors they make. Use the words "science experiment" when you talk about what they are doing. After a while, lay out sheets that are secondary colors so that your students can compare them with the combined primary color sheets.

Science kits: Make experiment kits for independent, individual exploration. Collect small gift boxes with lids (or stationery boxes) and put a collection of acetate color shapes in each one. Lay them out for the children to take and use as they please.

Secondary Rhythm Chant
Music (Rhythm)/Cognitive

WHY we are saying this chant: to reinforce rhythmically the concept of secondary colors, and to help children feel rhythm.

WHAT we will need:
  Rhythm sticks
  Chant:

  "Secondary, secondary,
  tell me what it means.
  Secondary's purple and it's orange
  and it's green."

HOW we will do it: tap the rhythm sticks on the floor in time to the chant. Experiment. Tap and chant softly, loudly, quickly, slowly. See if you all stop at the same time. If possible, provide rhythm sticks that are purple, orange, and green.

Color Breakdown
Science

WHY we are doing this project: to reinforce the concept of secondary colors in reverse— that is, by enabling the children to separate them into the original primary colors that made them.

WHAT we will need:
  Coffee filters
  Markers (orange, black, purple, green)
  Saucers or very shallow containers
  Water
  Book: Aliki, *How a Book Is Made*

HOW we will do it: To prepare, cut the coffee filters into single strips about 3 cm x 8 cm (1" x 3½"). On the activity table, provide small containers of water and the orange, green, purple, and black markers. (You may want to use old markers. Sometimes during the exploration, children dip the markers in water and the pens have to be discarded after the activity because they no longer write.) Color one end of the filter strip. Dip the colored end into the water, leaving the rest of the paper on the edge of the container. Watch the color move and separate. Do this experiment yourself, first. Some markers contain ink that doesn't separate. My experience is that even if the colors do not separate, children are fascinated with how the color moves and bleeds from one end of the filter strip all the way to the other, but obviously the lesson learned is a different one.

Read *How a Book Is Made* to the children, before beginning the activity. With three- and young four-year-olds, talk them through the illustrations. Pages 18 to 24 show how the color of each picture is separated into yellow, blue, red, and black. Spend extra time discussing these pages, and then let the children know they can do their own color separation. Show them a filter strip with a patch of marker color near one end. Show them a saucer of water. Ask them to predict what will happen if they dip the colored end in the water.

Invite the children to explore the materials on the activity table. Encourage them to verbalize the results of their science experiment. (Green breaks down into blue and yellow; black breaks down into many colors; purple breaks down into red and blue; orange breaks down into yellow and red.)

Dye Delight
Science/Language

WHY we are doing this project: to facilitate an understanding of dyes; to show children through hands-on, child-initiated activity that foods contain colors which can be used as dyes; to develop an appreciation for literature by reading *The Goat in the Rug;* to promote self-esteem by enabling children to make something they can keep.

WHAT we will need:
  White sheet (cut into handkerchief
    squares) or: plain white handkerchiefs
    (about 12.5 cm x 12.5 cm [5" x 5"] but
    can be larger)
  Five large-sized beets
  Four large-sized onions
  Plastic margarine containers
    (one for each child)
  White, sticky labels (one for each child)
  Tablespoons (one for each child)
  Clothes rack or clothesline (use yarn if
    necessary)
  Newspaper
  Large basin
  Book: Blood, Charles L. & Link, Martin,
    The Goat in the Rug

HOW we will do it: Set aside one raw onion and beet. Boil four beets in eight cups of water

to make purple water and the skins of four onions in the same amount of water to make brown. Save the cooked beets and onions. Also, set one handkerchief aside.

Read *The Goat in the Rug* before beginning the activity. Spend extra time on the pages which describe and show the wool being dyed. Show the children your raw beet and raw onion. Then show them the cooked vegetables, and the liquid they produced. Ask the children what they notice about the liquids. Hold up a handkerchief and ask the children to predict what will happen if they soak it in one of the liquids. When the children move to the activity table, invite them to use the tablespoons to ladle one of the liquids into their plastic margarine containers, and if you are working with three- and young four-year-olds, help them with this step as necessary. Use the sticky labels to put a name on each container. Encourage your students to put their handkerchiefs in their containers, and to let them soak overnight. For even dyeing, make sure all of each handkerchief is immersed in the dye. The next day, invite the children to squeeze their handkerchiefs into the basin, and to hang them on the clothesline or rack. How have the handkerchiefs changed? Are there any surprises? (The brown onion skin liquid produces a yellow dye.) Let the children compare them to the undyed handkerchief you set aside. Have new name labels ready. Some sticky labels stick onto damp fabric. If yours do not, stick the labels onto the line or rack, next to the appropriate handkerchief. When the handkerchiefs are dry, have a show-and-tell time if you like. Encourage the children to describe how they dyed their handkerchiefs, and to say what they will use them for. For an interesting variation, twist handkerchiefs tightly, secure with clothes pins, and then leave them overnight in the dyes. What result do the children observe?

Candy Colors
Science

WHY we are doing this experiment: to show that dyes are used to color food and to help children distinguish between natural and artificial colors.

WHAT we will need:
- M&M candies
- Small, shallow dishes
- Eye or medicine droppers
- Cotton swabs
- Water
- Bottles of food coloring
- Tray
- Soapy water
- Paper towels
- Green leaf
- Flowers of various colors
- Red tomato or apple
- Yellow squash
- Brown potato
- Any other natural object

HOW we will do it: On your activity table, arrange the shallow containers, cotton swabs, candies, water, and eye or medicine droppers. If you work with three- and young four-year-olds, your children may prefer to eat all the candies rather than experiment with them, so use your own judgment about doing this activity.

Arrange all your flowers and/or vegetables on the tray and have it nearby. During an Attention Getter time, pass around several candies for each child to eat. After the chocolates are eaten, ask the children to look at their hands. What do they see? Hold your own hand up so that the children can also see the color which came off on your skin and say, "The color of this candy is *artificial*. People added a dye to the candy part around the chocolate." Say the word *artificial* several times and with your fingers rub the candies as you say the syllables in order to make more dye come off. Show the children the bottles of food coloring and, together, as a group, identify each color. Explain that colors

like the ones in the bottles are often added to foods. Use the soapy water and paper towels for cleaning hands.

Next, encourage the children to touch and handle the objects on the tray. Ask them to notice what their hands look like after handling the fruits, leaves, and vegetables compared to how their skin looked after handling the candies. (Did any color come off?) Ask the children to hypothesize about why this is so, and, during the discussion, use the words *artificial* color and *natural* color. Have a discussion about the children's hair colors and whether they are natural or artificial.

Explain to the children that if they would like to experiment more with artificial colors, they can drop water drops on the candies to see what happens. Ask them what they see in the room that will help them conduct this experiment. When they discover the materials on the activity table, allow the children to explore them freely. As water is added to the candies, what happens to the dye colors? Cotton swabs can be used for mixing the colored water and making dye combinations. What happens when the same experiment is conducted on the objects from nature?

Dyeing Food
Science

WHY we are doing this activity: to enable children to use food coloring to dye their own food, and to provide more experience with mixing colors.

WHAT we will need:
- Scrambled eggs or vanilla pudding (not too yellow)
- Milk
- Plastic forks
- Food coloring
- Small containers
- Cups
- Plates
- Spoons or forks

HOW we will do it: If you want this project to be another experiment in mixing primary colors, then provide only yellow, red, and blue food coloring and only milk; otherwise provide other ingredients also.

Arrange all materials on the activity table. Make your scrambled eggs or pudding together as a group project, or you may choose to use instant pudding. Ask the children to predict what they will see if they add each food coloring to the foods. Help the children serve themselves. Then encourage them to dip their forks in the food coloring and to let the drops fall onto their food. (Eye and medicine droppers can pull up quite a bit of liquid, and food coloring is expensive. By using forks you ensure that the children do not use too much; however, if you work with kindergartners you may choose to use eyedroppers instead. If you do this, make sure they are thoroughly sterilized by soaking them in either weak bleach water or hot, soapy water and then rinsing well.) As the children experiment, ask: "What's happening?" Invite your students to compare what they see when a color is dropped into milk, with what they see when the same color is dropped onto scrambled eggs or yellow vanilla pudding. When the children eat their colorful snacks, does the food taste different?

Cabbage Colors
Science

WHY we are doing this project: to help children understand that chemical reactions can produce color changes; to develop all components of language arts: reading, writing, speaking, and listening; to develop fine motor skills through pouring and stirring.

The facts of the matter: Cabbage juice and other juices contain a chemical called an *indicator*, which causes the juice to turn different colors when it is mixed with an acidlike vinegar or lemon juice, or a base, like soap or baking soda. In general, bases are bitter and tart, whereas acids are sour.

Developmental differences: Young children will not absorb the above information. If you are working with kindergartners, you can introduce these facts as and how you think appropriate. The important thing is that this hands-on, child-initiated experiment will lay a foundation for later knowledge; for now, making the cabbage juice turn green, red, or pink is just plain fun. Also, three- and young four-year-olds will not use the science experiment writing sheets for filling in their results, but they will scribble on them, and often they will recognize the symbols on the sheet as being associated with the project they just participated in. It is helpful to children to refer to their scribbles as "writing" when talking with them about their work.

WHAT we will need:
 Purple cabbage
 Vinegar
 Liquid soap
 Baking soda
 Water
 Plastic margarine containers
 Plastic apple sauce containers
 Spoons or medicine droppers
 Sensory table or tubs
 Small pitchers
 Blank paper
 Pens
 Crayons or markers (pink, red, and green)
 Activity sign and writing sheets (format provided on page 170; photocopy and enlarge for your use)

HOW we will do it: To prepare, boil purple cabbage until you have enough cabbage juice for the project. Set out the plastic margarine containers in the tubs or sensory table. Nearby, place smaller, separate containers of vinegar, baking soda, water, and liquid soap. Put medicine droppers or spoons in them. Have several extra spoons or droppers available. Put the cabbage juice in the small pitchers and place them in the tubs or sensory table. Make photocopies of the writing sheet in the text, and on a table nearby, place these, blank sheets, and pens, crayons, or markers. Make

WHAT HAPPENS WHEN YOU MIX CABBAGE JUICE WITH BAKING SODA VINEGAR AND LIQUID SOAP ??

WHAT COLOR??

CABBAGE JUICE + BAKING SODA = _____

CABBAGE JUICE + VINEGAR = _____

CABBAGE JUICE + LIQUID SOAP = _____

the activity sign, and hang it up beside the tubs or sensory table.

Before the children approach the project, help them interpret/read the activity sign and invite them to carry out its suggestions. Ask them to predict what will happen when they conduct the experiment. As the students explore the materials, discuss their results. What happens to the cabbage juice when the other materials are added to it? Invite the children to use blank paper or the science experiment writing sheets to record their results. Older children often invent their own spelling or ask to have words written down that they can copy onto the sheets, and younger children like to give dictation. When they are finished, ask the children to read their experiment results to you, or if they prefer, read the record of their experiments to them. Other ways to develop speaking and listening skills are to have a show-and-tell of experiment result sheets after the activity, a group sharing time about the experience, or use a puppet while the children are actually experimenting and have the puppet ask them questions about what they are doing. (This last method is really fun, and elicits *a lot* of language!)

Prism Project
Science

WHY we are doing this project: to help children understand that ordinary light is made up of a combination of colors; to show children that a prism breaks down these colors; to develop a sense of autonomy through use of a one-person work station.

The facts of the matter: The different colors in a beam of light travel at the same speed in air, so they stay combined and are seen as white light. When light travels through glass, however, the various colors travel at different speeds and separate from each other. One color travels faster or slower than another, and in this separation, the colors can be seen individually. The best analogy for this is a race. People in a race do not all run in the same spot at the same pace; they run at different speeds and spread out from each other.

WHAT we will need:
Prism (available at nature stores, teacher supply stores, or toy stores)
Flashlight
Cardboard (18 cm x 12.5 cm [7" x 5"])
Sturdy block
Sheet of white construction paper
Small table and one chair
"One person may be here" sign (format provided on page 165; photocopy and enlarge for your use)
Blue tack or putty

HOW we will do it: To prepare, punch a small hole in the center of the cardboard, and using blue tack or putty, secure the cardboard to the block so that it stands by itself. Hang the sheet of paper on a wall, or tape it to a book and spread the front and back of the book apart so that it is freestanding, or use a typist's stand.

Align all materials on the table so that the cardboard is nearest the chair, the paper is farthest away, and the prism is on the table, in the middle.

Make sure that when the flashlight is shining though the hole in the cardboard, it hits the prism. Ahead of time, discuss the "One person may be here" sign with the children. Ask the children to predict what they will see when they turn the flashlight on and shine it through the hole in the cardboard. As they explore the materials, talk to them about what they see on the paper (the color spectrum). After the day's activities, have a group discussion about this science experiment and what it tells us about light, as well as what it tells us about prisms. (See "The facts of the matter.")

Making Rainbows

Science

WHY we are doing this project: to reinforce the concept of the color spectrum.

The facts of the matter: A rainbow is formed by the reflection, refraction, and dispersion of sun rays in falling rain or mist. Refraction is the bending of a ray or wave of light.

WHAT we will need:
> Spray bottles on "mist" setting or hose with sprayer attachment (if outdoors)
> Water
> Hot, sunny day

HOW we will do it: Invite the children to spray a fine mist in the sunshine. What do they see? (They should be able to spot a rainbow.) During discussion, use the words *color spectrum* and *arc*.

Developmental differences: Three- and young four-year-olds enjoy the sensory aspect of this activity. The power of having their very own spray bottles to activate excites them and they may be most interested in spraying each other, which is fine on a hot day. Older children will be more inclined to observe and discuss the rainbows they make, and may also be inclined to spray each other.

Color Spinners

Science/Fine Motor

WHY we are doing this project: to introduce children to the idea of an optical illusion involving color (when the spinner spins, the color dots seem to merge and spread into circles); to develop self-esteem by enabling children to make something they can keep; to develop fine motor skills.

WHAT we will need:
> Cardboard (cut into circles that are 7.5 cm [3"] in diameter)
> Small pencils
> Markers or crayons (red, green, and blue)

HOW we will do it: To prepare, use a skewer to make a hole in the middle of each cardboard circle. Make a sample spinner by using the markers to randomly place red, green, and blue dots all over the circle. Next, push a pencil through the middle, point end down. When the children are gathered, give them the opportunity to closely examine your spinner and how it is made. Ask the children to predict what they will see when the spinner is spun, but do not actually spin it. (The children may spin it themselves though, in the course of examining it, and this is fine.) Ask them what they see in the room which will help them make their own spinners. Invite them to find out if their predictions are right by making spinners and then twirling them. As the children twirl their spinners, ask them what they see. Encourage them to compare what they observe when the spinners are moving to what the spinners look like when they are still.

Note: Some parents and teachers do not like to use adult samples when introducing a project, but my experience is that when the purpose of materials is not immediately or clearly evident, samples generate much more interest in the activity. My solution is to let the children know that my spinner (or collage or sculpture or whatever) looks the way *I* wanted it to, and that each person's will look different because each person is different, and that we all make things in our own special way. Also, it is a good idea to put the sample away before the children begin their exploration of materials.

Developmental differences: Three- and young four-year-olds sometimes have a hard time spinning spinners, and get very frustrated. You know best what level of fine motor development your children have reached, and accordingly, you can decide whether or not to do this project.

Color Spots
Gross Motor

WHY we are doing this project: to facilitate color recognition and to develop the large muscle group.

WHAT we will need:

Construction paper to cut out large circles
Contact paper

HOW we will do it: To prepare, take a large plate and trace circles onto a wide variety of differently colored construction paper sheets, making sure you have several of each color. Cover with contact and spread the shapes out over a large, clear space. Gather the children together and say: "Color a little, color a lot, can you find a big red spot?" (For colors with two syllables, leave out the word *big* when you say the rhyme.) As soon as you say the rhyme, the children can run to a spot which is that color and touch it. More than one person can touch a spot. After the children learn the game, let them each take a turn to say the rhyme and choose the color. Another variation of this game is to put the color circles in interesting, out-of-the-way places to see if the children can find them: under a chair, on a toy shelf, in the block bin, and so forth.

    Tip: The color circles will be even sturdier if you laminate them instead of using contact paper.

Eggsactly
Math

WHY we are doing this project: to practice rational counting; to facilitate color matching; to reinforce the names of colors for younger children; to develop reading and writing skills for older children; to develop self-esteem and a sense of autonomy through use of a one-person work station.

WHAT we will need:

Large plastic eggs of various colors
    (available at craft, hobby, and toy stores)
Small objects of matching colors (tiddly
    winks, small Legos, figurines, etc.)
Tray

Older children:

Small pencils
Blank paper
Writing sheets (format provided on
    page 174; photocopy and enlarge
    for your use)
"One person may be here" sign
    (provided on page 165; photocopy and
    enlarge for your use)

HOW we will do it: To prepare, gather together the above items. If you are doing this project with older children, and you have no small pencils, ask your local library for a donation and explain that you want them for the children you teach. (Most libraries provide small pencils near the card catalogs so that patrons can write down reference numbers.) You may want to have as few as four differently colored eggs on the tray if you are short of time, because for each color of egg you have, you will need to make several writing sheets. Make photocopies of the sample writing sheet, making sure you have plenty of spares. Put a scribble of the appropriate color above the color word on each sheet. Cut out blank paper which is the same size as the writing sheets and make these available also, for children who want to create their own writing sheets.

    Next, put a variety of objects inside each egg, making sure that they correspond in color. For older children, put a small pencil and writing sheet or blank sheet inside as well. Put all the eggs out on a tray. Gather the children together and show them the eggs. Take one and shake it, and ask them to guess what is inside. Open it, and together, count the number of objects inside. If you are teaching older children, show them the pencil and writing sheet or blank paper, and write the number you counted on the paper. Also, show them the "One person may be here" sign, and discuss what it means. Replace the writing sheets and blank paper as needed.

How many **LEGOS** in
the **PINK** Egg?

_____

Here's a sample writing sheet.

Draw pictures that represent the objects
you put in your eggs. Write the name (word)
underneath. Photocopy these and put a
scribble of the appropriate color above the
written word which names the color.

How many _____ in
the _____ Egg?

How many _____ in
the _____ Egg?

How many _____ in
the _____ Egg?

How many _____ in
the _____ Egg?

174

Developmental differences: Three- and young four-year-olds may have more fun shaking the eggs, opening and closing them, and engaging in pretend and manipulative play with the objects inside. Even if they choose not to count the objects, the concept of *things of like color* will be reinforced. Older children will be more likely to count all the objects and to write down the number counted.

Color Match
Math

WHY we are doing this project: to facilitate color matching; to reinforce the names of colors for small children; to facilitate dual matching for older children; to facilitate cooperation and communication through use of a two-person work station.

WHAT we will need:
 Pattern (provided on page 176; photocopy and enlarge for your use)
 Stiff board (cardboard or foam core)
 Clear contact paper
 Colored construction paper
 Double-sided tape
 Black marker
 "Two people may be here" sign (provided below; photocopy and enlarge for your use)

HOW we will do it: Foam core consists of a thin sheet of styrofoam sandwiched between two pieces of stiff paper and can be purchased at hobby, craft, and teacher supply stores. To prepare this activity, use the pattern to cut a car and garage out of each color of construction paper. If you are teaching young children, this project is a simple matching project and you will only need one of each color. For older children, make the game more challenging by also drawing stripes or dots on the cars and

2
Two people may be here.

176

garages, for more complex matching. Glue the garages onto the board, and cover the board with clear contact paper. Cover each car on both sides with contact paper. Place double-sided tape onto each car or each garage, so that as the children match the cars to the garages, they stick onto the board.

Before the children use the materials, discuss the meaning of the "Two people may be here" sign. With younger children, use the names of the colors as you talk to them about what they are doing.

A Book of Many Colors
Language

WHY we are doing this project: to develop all components of language arts and to reinforce understanding of different colors.

WHAT we will need:
    Magazines with color photos
    Children's scissors
    Pale-colored construction paper
    Glue
    Crayons
    Stapler

HOW we will do it: To prepare, make a blank book for each child. To do this, cut construction paper into pages that are approximately 30 cm x 45 cm (12" x 18"). For each book, put one sheet over another, and fold in half. Then staple twice along the fold.

Leaf through your magazines and tear out pages with the most colorful and interesting pictures. Older children (older four- and five-year-olds) may want to cut out their own pictures, but for younger children, it is a good idea to cut out most of the pictures yourself. Leave some scissors and pages on the table as well, so that cutting is an option. In addition to the magazine pictures, set out the blank books, glue, and crayons on the activity table.

This is one of those activities in which the children show much more interest when a sample book is read to them first. Make a sample

using photographs with good, strong color and make sure you include pictures that feature black and shades of brown. You can title your book *A Book of Many Colors* or create your own title. Print a sentence or two about each picture, for example, "I like this picture of an alligator because the green is the same color as my sweater and it's my favorite color." "This brown chocolate bar looks so good." "I picked this picture to go in my book because the purple violets are so pretty." Read the title and pages of the book to the children and encourage their comments. Ask the children what they see in the room that would help them make their own color books.

When the children have glued their pictures onto the pages, take story dictation if appropriate. Older children may want you to write their words down on a separate piece of paper so that they can copy them into their books, or they may invent their own spelling. When the books are finished, ask the children to read them to you, or if they prefer, read their books to them, or have group time show-and-tell. If you choose to have show- and-tell, have a small number of children show- and-tell their books each day, over a week or two.

Color Circle
Language/Small Group Activity

WHY we are doing this activity: to develop speaking and listening skills and to expand vocabulary.

WHAT we will need:
    Construction paper shapes (one of each
      of the eight basic colors)
    Large box or bag

HOW we will do it: To prepare, gather together your construction paper shapes. If you like, use the ones that you made for the previous Color Spots gross motor exercise. Put the shapes in your box or bag, and have the children sit in a circle. Tell them that you are going to pull a color out of the box, and that everyone who is wearing that color can stand up. Tell the children that they can also be "wearing" the color of their hair or their eyes. After you pull out a paper color shape, and those children are standing, ask each one to tell you about whatever they are wearing that matches in color. Encourage the children who are still sitting to listen to the answers. Be sure you pull out colors which give all children a chance to stand and tell about the colored item they are wearing. After a while, let the children take turns at pulling the paper color shapes out of the box and asking the questions.

Blurry Colors
Art/Cognitive

WHY we are doing this project: to facilitate creative expression and to reinforce, through art, the fact that primary colors are red, yellow, and blue, and that they make secondary colors when they are mixed.

WHAT we will need:
    Paint (primary colors—red, yellow, blue)
    Pale-colored construction paper
    Shallow pans
    Water
    Fine paintbrushes
    Containers
    Newspapers

HOW we will do it: To prepare, spread several layers of newspaper on your activity table. Mix up your paints and put them in the containers. Pour water in the shallow pans. Place all materials on the activity table.

When the children approach the table, ask them to dip a piece of paper into a pan of water, and then to hold it over the pan while the excess water runs off. Then invite the children to paint on the paper. As the primary colors blur and run together, what happens? Talk about the new colors that are made. As appropriate, reinforce the fact that the colors the children started with are primary, and the new colors that are created are secondary.

Spangles and Sequins
Art

WHY we are doing this project: to develop fine motor skills and to facilitate creative expression.

Children love using spangles and sequins for art projects, and because they come in so many vibrant colors, they are perfect for this unit. Also, they are small and are just right for fine motor development. Hobby and craft stores often sell them, or they can be ordered from teacher supply catalogs.

WHAT we will need:
    Spangles and sequins
    Glue
    Glue brushes
    Paper
    Shallow containers
    Cardboard

HOW we will do it: To prepare, pour spangles and sequins into the containers, and arrange the containers so that all children will be able to reach them easily. You do not have to use cardboard for this project, but because it is so sturdy, it lasts longer. Cut the cardboard into rectangles of 12.5 cm by 15 cm (5" x 6"). Set them out on the activity table with the spangles, sequins, glue, and glue brushes. As the children make their creations, talk about the different colors they are choosing. This is also a good opportunity to introduce the colors gold and silver. With older children, talk about which sequins and spangles are primary colors and which are secondary.

Math option: Invite older children to count how many spangles or sequins they used, and to write the number onto their cardboard pieces, or to make patterns.

Colorful Masks

Art/Multicultural

WHY we are doing this project: to facilitate creative expression; to reinforce the names of colors; to introduce children to a French custom.

The facts of the matter: Mardi Gras is held on Shrove Tuesday, the day before Lent begins, to mark the end of a long carnival period which begins on January 6. It was brought to America by French colonists in the early 1700s. Mardi Gras means "Fat Tuesday" and may refer to the custom of parading a fat ox through French villages and towns on Shrove Tuesday. Today, the New Orleans celebration of Mardi Gras is the most famous. People parade in colorful masks and fancy dress through the streets of the city. Societies called *krewes* organize and pay for festivities and parades. During the carnival season, *krewes* also give balls and parties.

WHAT we will need:
- Book: *Mardi Gras*
- Mask pattern (provided on page 180; photocopy and enlarge for your use)
- Popsicle sticks or tongue depressors
- Construction paper
- Acetate paper cut into strips
- Spangles and sequins
- Yarn or string
- Feathers (real or fake)
- Glitter or colored sand
- Any other interesting materials you have on hand

HOW we will do it: To prepare, use the mask pattern to cut out masks from different colors of construction paper. An easy way to cut the eyes and noses is to fold the mask first, as shown in the illustration, or use an Exacto knife. If you want to use real feathers, ask pet stores to save the feathers molted by birds. Avoid feathers from wild birds, as they may carry disease. Set all materials out on the activity table.

Make a sample mask by picking your favorite color of paper and decorating it with the materials. Try to put something of every color of paper on your mask. Then glue a Popsicle stick or tongue depressor onto it, so that you can hold your mask up. You may want to reinforce the back of the masks with extra Popsicle sticks so they are not too floppy to hold up. Make your sample at least two days before the children make theirs, so that the glue has time to dry. With the children, look at the photographs in the book *Mardi Gras*.

During an Attention Getter time, hold your mask up in front of your face, and tell the children that you picked your favorite color of paper and made a mask out of it. Point to the objects you glued on your mask, and as a group, say what color each one is. Tell the children that you put something of every color on your mask. Hold up the picture of Mardi Gras and use the above facts as you think appropriate to discuss this custom. Ask: "What do you think the people are doing? What are they wearing? How are they covering their faces? Which costume in the picture is your favorite and why?"

Encourage the children to make their own masks, and comment on the colors they choose as they work on their creations.

Mardi Gras

Dramatic Play/Language/Multicultural

WHY we are doing this: to reinforce names of colors; to develop speaking and listening skills; to facilitate self-expression; to facilitate social interaction; to help develop coordination of actions and words; to introduce children to a French custom.

WHAT we will need:
- Colorful masks (made in previous activity)
- Dress-up clothes (especially long jackets, skirts, and dresses)
- Hats
- Jewelry
- Music
- Children's full-length mirror

HOW we will do it: To prepare, set out the dress-up clothes, hats, jewelry, and mirror. If you have trouble collecting dress-up clothes, ask friends, parents, or neighbors for donations of old garments. Invite the children to dress up

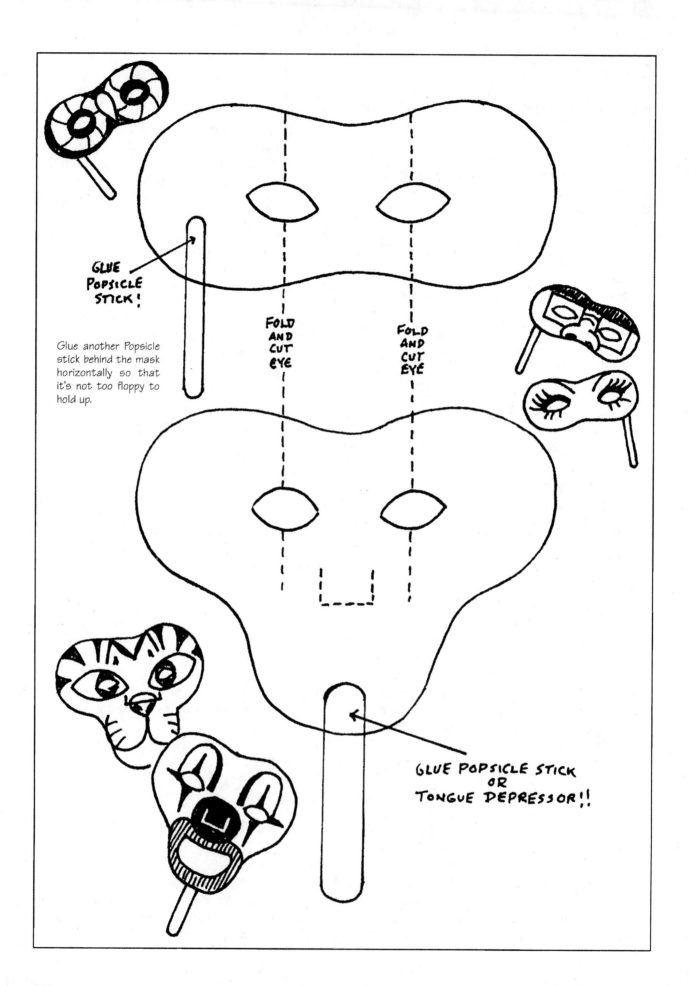

GLUE
POPSICLE
STICK!

Glue another Popsicle stick behind the mask horizontally so that it's not too floppy to hold up.

FOLD
AND
CUT
EYE

FOLD
AND
CUT
EYE

GLUE POPSICLE STICK
OR
TONGUE DEPRESSOR!!

180

for their own Mardi Gras carnival, using the masks they made themselves. You may even want to dress up yourself, to help get them in the mood. Talk about the colors each child is wearing. Give the children time to explore this dramatic play, and ask them if they would like to parade through your home, neighborhood, school, or classroom.

Skin Colors
Multicultural/Social Studies

WHY we are doing this project: to help develop appreciation of the differences between people.

WHAT we will need:
Book: Mandelbaum, Pili, *You Be Me, I'll Be You*
Flour
Instant coffee
Water
Tub of water
Paper towels
Mirror

HOW we will do it: To prepare, make a thin, paint-like substance with instant coffee and water. Set this out with the flour and the mirror. During an Attention Getter time, read *You Be Me, I'll Be You* and talk about the story. Afterward, have everyone hold one bare arm out, and put your arms next to each other. Talk about the different colors of skin you see: light brown, pink, freckled, dark brown, and so forth. Show the children the coffee liquid and the flour, and invite them to change their skin color by dabbing and painting the colors on their skin. The tub of water and paper towels can be used for clean-up time. Children enjoy this activity immensely.

Literature

Symbol Key: * Multicultural
              + Minimal diversity
              No symbol: no diversity or no people

Coil, S. (1994). *Mardi Gras*. New York: Macmillan.

Dodds, D. A. (1992). *The color box*. New York: Little, Brown & Co.

Emberley, E. (1992). *Go away big green monster!*. New York: Little, Brown & Co. (The first three books in particular are terrific. The writers have utilized shapes cut out of the pages to produce ingeniously designed books. Children love them.)

Hoban, T. (1978). *Is it red? Is it yellow? Is it blue?*. New York: Greenwillow Books.*

Hoban, T. (1989). *Of colors and things*. New York: Greenwillow Books.

Jonas, A. (1989). *Color dance*. New York: Greenwillow Books.

Lionni, L. (1985). *Colors to talk about*. New York: Pantheon.

Lionni, L. (1959). *Little blue and little yellow*. New York: Astor.*

Mack, J. (Ed.). (1994). *Masks and the art of expression*. New York: Abrams. (For mask project and dramatic play:)

Mandelbaum, P. (1990). *You be me, I'll be you.* Brooklyn, New York: Kane/Miller.*

Rogow, Z. (1988). *Oranges.* New York: Orchard Books.*

Extenders

Music and Movement: Read *Color Dance* by Ann Jonas, and then let the children dance to music with differently colored scarves.

Cognitive: Distribute a variety of colors of construction paper shapes to each child, and chant:

> Color criminy color cram,
> Hold a green one in your hand.

Change the color each time you chant.

When the children understand the game, let them take turns being the one to say the rhyme and to choose the color that the other children have to hold up. With older students, say:

> Color criminy color cram,
> Let's see primary in your hand!
> or: . . .How 'bout secondary in your hand?

Math: Save small, interesting boxes (children especially love boxes with lids that snap) and on the lid of each, tape or glue a colored piece of acetate paper. Inside each box, put several pieces of the matching color of acetate, or paper cut from plastic or cardboard folders, and invite the children to count how many of each color are inside each box.

Science: After your experiments with food and food dyes, make a display from dyed foods found in the supermarket. Some ideas might include: pickles, Cheetos, colored cereal, and sweet and sour sauce. Talk about the ingredients of these foods and compare their nutritive value.

SOAP SCIENCE

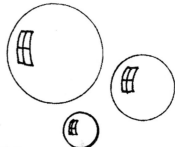

Attention Getter: A fun way to begin this unit is to invest in a few shaped soaps (animals, seashells, flowers, etc.), and let the children examine them. If you do not have any shaped soaps, kick off the activities by asking the children to close their eyes and smell a bar of soap. Can they guess what it is? Or bring in a selection of different bar, liquid, and powder soaps to discuss and compare. Ask the children if they can guess what you will be working with over the next few days. A safety precaution: remind the children before every activity that it is important not to touch or rub their eyes while using soap.

Comparing Soaps
Science/Sensory/Language

WHY we are doing this project: to facilitate scientific observation of soap in its different forms; to provide children with a sensory experience; to develop speaking, listening, reading, and writing skills.

WHAT we will need:
    Liquid dish soap
    Soap leaves
    Powdered laundry soap
    Ivory Snow flakes
    Bar soap
    Pitchers of water (warm and cold)
    Tub of cold water
    Forks
    Spoons
    Funnels
    Aquarium tubing (available at pet stores)
    Small containers
    Newspaper
    Tubs or sensory table
    Paper towels
    Butcher paper
    Markers

HOW we will do it: To avoid a catastrophic mess, put small amounts of each kind of soap into the containers. Spread plenty of newspaper under your work area. Pin the butcher paper up on the wall at the children's level. Put all the materials in the sensory table and encourage the children to explore them.

As they touch and use the different soaps, ask them how each one feels. If a discussion is slow to get started, introduce some descriptive words into the conversation; *gritty*, *slimy*, and *soft*. Encourage observations. At the top of the butcher paper, write: "How do different soaps feel and look?" If you work with younger children, write down their comments. Use quotation marks and write their names next to their words. Use alternate colors of markers for each child's comment so your students can distinguish individual sentences. If you work with older children, invite them to write their own comments and/or draw pictures on the chart. At the end of the session, read the chart together.

Making Bubbles
Science/Math

WHY we are doing this project: to facilitate cooperation as a group and to practice rational counting with cup measurements.

WHAT we will need:
    1 cup measuring cup
    ½ cup measuring cup
    Large container
    6 cups water
    Joy liquid detergent (2½ cups)
    3 cups glycerine (available at drugstores
      and some supermarkets)
    Large spoons
    Wire bubble blowers (instructions below)

HOW we will do this: Ahead of time, twist wire into bubble blowers by wrapping each piece around a cup to form a round circle. Make this bubble liquid together as a group, after the introductory Attention Getter discussion. Tell the children how many cups of each ingredient you need, and have them count each cup you pour into the container. If possible, have the children take turns pouring a cup of an ingredient into the mixture. Let them use the large spoons to mix everything together. Pass around the bubble blowers and invite the children to dip them into the bubble mixture and blow.

Making a Bubble Holder
Science/Fine Motor/Crafts

WHY we are doing this project: to develop the fine motor muscle group; to enable children to examine bubbles closely.

There are several ways to make a bubble holder, so choose the method which utilizes the materials that are easiest for you to obtain. Or, if you like, make all the bubble holders and compare how well they work.

WHAT we will need:
    Plastic cups
    Spools
    Pencils
    Wire
    Small funnels
    Bottles
    Magnifying glasses
Preparation:
    Scissors
    Sharp knife

HOW we will do it: To make a bubble holder from a plastic cup, turn the cup upside down and use a sharp knife to punch a horizontal slit in the side of the cup, directly beneath the cup bottom. Then use scissors to cut the bottom out. Bubbles can be blown through the cup and then placed on the rim.

To make a bubble holder with a spool, insert a pencil in the hole of the spool, and wrap one end of a piece of wire around the pencil. Wrap the other end around a glass or round container to get the best circular shape you can, and then remove it. The spool will hold the pencil upright and the wire will remain horizontal. To use a funnel for a bubble holder, simply turn it wide side up, and put the narrow end in a bottle.

Make sure all the bubble holders are dipped into bubble liquid before transferring bubbles to them; dry holders will pop the bubbles. Blow bubbles through the cup, funnel, or wire and then set the holders down to examine the bubbles, or use a bubble wand and gently transfer the bubbles to soapy holders. Children can gently transfer bubbles to the holders, or catch bubbles with them. Show them the magnifying glasses and encourage them to use them to examine the bubbles closely.

Bubble Palace: It Takes Air to Make Bubbles
Science

WHY we are doing this project: to allow children to understand how the force of air is necessary to make bubbles.

WHAT we will need:
    Bubble liquid
    Straws
    Shallow containers
    Food coloring

HOW we will do it: Set out the containers and put a little bubble liquid and food coloring in each one. Place a straw next to each container. Invite the children to blow into the liquid through the straws and see what happens. If you work with three- and young four-year-olds, have the children practice blowing out through the straws instead of sucking in. Talk about what will happen if they suck up the soapy liquid through their straws.

Ask: "Do you think you can make a bubble without air or wind?" Invite the children to try. When they begin to use the straws to make bubbles, invite them to see how high they can build up their bubble palaces. Explain that the children's breath is forced through the straw and into the soapy liquid, where the air and the force combine to make bubbles.

More Proof That It Takes Air to Make Bubbles

Science/Gross Motor/Cognition

WHY we are doing this project: to provide children with another method to discover that it takes air and force to make bubbles; to develop large muscles; to help children understand how the gears in beaters work.

WHAT we will need:

Manual rotary eggbeaters
Whisks
Forks
Large slotted spoons
Tubs or sensory table (filled with water)
Liquid soap
Small containers (for liquid soap)
Eyedroppers
Spoons
Newspaper
Large pitchers of cold water
Large empty tub

HOW we will do it: To prepare, spread several layers of newspaper under your work area. Set out your tubs of water with the manual rotary beaters and whisks. Use the large pitchers of water for refills, and the large empty tub for dumping soapy water. Young children can often have great difficulty operating manual rotary beaters because they are large and cumbersome, so use your own judgment about whether or not to provide these with the other materials.

During an Attention Getter time, ask them to predict what will happen if they mix the water with soap by using the beaters, whisks, and slotted spoons. Encourage them to conduct this experiment. Remind the children of what they learned when they used the straws with soapy water in the previous experiment and, using this information, encourage your students to hypothesize about why beating or whisking the soap and water results in bubbles. (The motion of the beaters forces air into the soapy liquid, and this makes bubbles.) Compare the bubbles made with the beaters to the bubbles made with the straws. Do they look the same? Why? (Beaters force more air into the soapy water and make many more bubbles. This is why the bubbles appear to be much frothier and foamier.)

At the end of the day or the session, talk about your findings. What did the children notice about which tool made the most bubbles the fastest? Demonstrate the manual rotary beater and discuss what makes the beaters turn around.

Bubble-Making Song

Music/Movement/Cognitive

WHY we are doing this activity: to musically reinforce the process involved with making bubbles; to help children enjoy using their singing voices; to develop cognition through memorization of words and related actions.

WHAT we will need:

Song: "BUBBLE-MAKING SONG"
(to the tune of "The Grand Old Duke of York")

"BUBBLE-MAKING SONG"
I take my bubble wand
and dip it in my soap,
wind is blowing,
bubble's growing,
and away it floats.
First it floats up high,
then it floats down low,
finally that old bubble bursts;
I wonder where it goes!

For "I take my bubble wand and dip it in my soap," pretend to be holding a bubble wand and dipping it into liquid soap. For "wind is blowing, bubble growing," blow air with your lips and then use your hands to simulate a growing bubble. For "First it floats up high, then it floats down low," cup your hands to simulate a bubble, and move your hands slowly up and then slowly down. For "finally that old bubble bursts," make a tight fist, and then suddenly open all your fingers wide. For "I wonder where it goes!" put your hands out in a questioning gesture.

Sing the song and do the motions until the children have learned them. When the song is very familiar to the children, try a variation of it. Agree to leave out the word "bubble," "float," or both when you sing the song. Can everyone remember not to sing those words?

Soap Boat
Science

WHY we are doing this project: to enable children to observe that soap breaks up the water molecules which create surface tension.

WHAT we will need:
  Half-gallon cardboard milk cartons (one for every four children)
  Boat pattern #1 (provided below; photocopy and enlarge for your use)
  Water
  Liquid soap
  Small containers (for liquid soap)
  Eyedroppers
  Tubs or sensory table (filled with water)
  Newspaper
Preparation:
  Skewer or needle
  Fingernail scissors
Demonstration:
  Two small containers
  Water
  Black pepper
  Liquid soap
  Magnifying glass
  Water refills

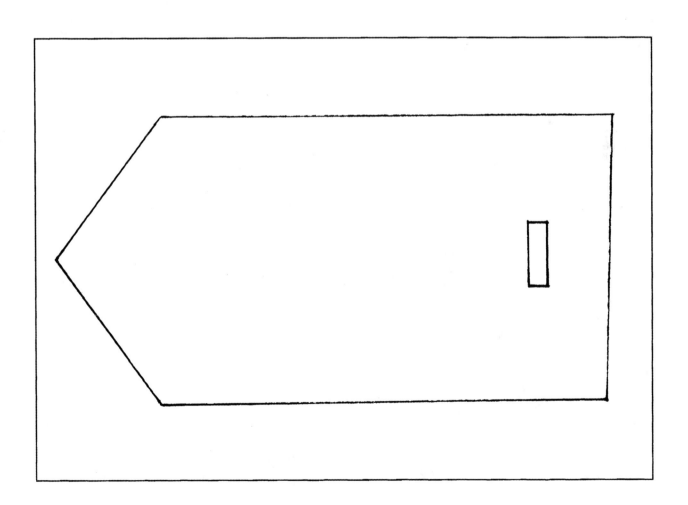

HOW we will do it: To prepare, use the provided boat pattern to cut flat, rectangular boats out of milk cartons. To make the rectangular hole, use a skewer or needle to poke a starter hole, and then use fingernail scissors to cut out the rest. This can be time-consuming, but children really get a kick out of watching the boats be propelled by soap power. Ask friends or parents to help you prepare one boat for each child.

On the day of your activity, spread out plenty of newspaper layers on the floor, under the project. Set out tubs of water or fill the sensory table. Put the boats, containers of liquid soap, and eyedroppers out on the activity table.

During an Attention Getter time, have the children lie on their stomachs in a circle, so that they can see your demonstration. Fill a small container with water and ask the children to predict what pepper will do when you sprinkle it into the water. After the children have made their predictions, sprinkle some pepper in your hand, and then drop it onto the water. What does the pepper do? Pass around the magnifying glass and encourage each child to notice how all the pepper grains sit on top of the water. Tell them that water has *surface tension* because the water molecules cling together. Ask the children if they have ever seen insects sitting on top of the water in a puddle or a pond. Insects can stand on water because of the surface tension. Drop several drops of liquid soap in the other container of water and stir in the soap with your finger. Ask the children to predict what will happen when you sprinkle pepper into this water, and then do so. What happens to almost all of the flakes? Why did they sink? Tell the children that the soap broke up the water molecules so that they spread apart and destroyed the surface tension. This is why the pepper sank instead of staying on top of the water.

Show the children the boats, and ask them to predict what will happen if they lay them on top of the water, and then use the eyedroppers to put drops of soap in the rectangular holes. Encourage the children to conduct this experiment. Remind them of the pepper experiment. Why do they think the boats move? If necessary, explain that when the soap breaks the water molecules apart, behind a boat, it pushes the boat forward. Encourage observations and discussion.

Note: In order for the experiment to work each time, everything must be washed free of detergent, and you must use fresh water. For this reason, if you do not have a sink in your activity room, use shallow amounts of water in your tubs to minimize the amount of refill water you have to keep on hand. Use one tub of clear water for children to wash their hands in when they want to try the experiment again.

Developmental differences: Three- and young four-year-olds may be more interested in floating and sinking the boats, splashing the water, and squeezing the eyedroppers. In the process of this sensory exploration, they are still likely to observe the effect of soap on the water and on the boats. Older children will be more likely to explore the materials with the specific goal of observing the soap's effects.

Another Soap Boat
Science

WHY we are doing this project: to enable children to experiment with another design of soap boat and to observe how it compares with the first boat design.

WHAT we will need:
    Half-gallon cardboard milk cartons
    Boat pattern #2 (provided on page 189;
        photocopy and enlarge for your use)
    Tubs of water
    Liquid soap detergent
    Small containers (for the detergent)
    Popsicle sticks

HOW we will do it: Use pattern #2 in the text to cut out flat, triangular boats from the milk carton sides. Fold the middle flap up so that the boat looks like the smaller diagram on the pattern. Fill the tubs with water, and put the detergent in the small containers with the Popsicle sticks beside it. Ask the children to predict if these soap boats will react to soap in the same way that the other ones did. Encourage the chil-

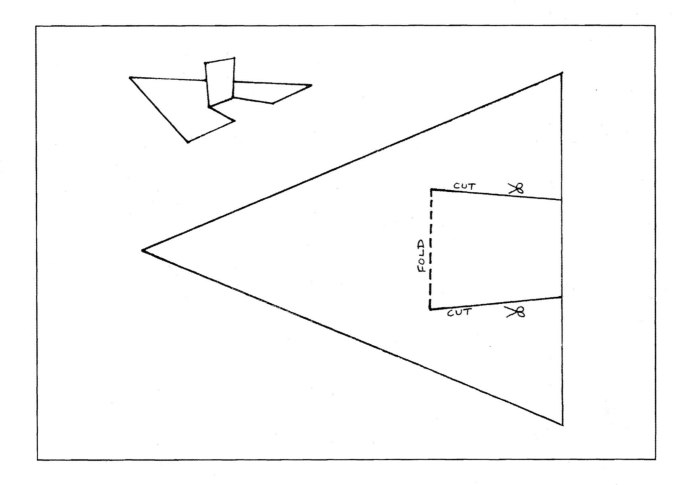

dren to set a boat on the water, and to touch the water space in the back of the boat with a Popsicle stick dipped in detergent. What happens? (The detergent reduces the surface tension in the back of the boat, so it is pulled along by the stronger surface tension in the front.)

Invite the children to put both types of boats side by side and to predict, and then compare, which one is pulled the farthest. To explore further, make both types of boats in different sizes, again, to compare which travels the farthest and the fastest.

It's a Shipwreck!
Science/Art

WHY we are doing this experiment: to provide children with a fun way to experiment with surface tension and soap's effect on it; to facilitate artistic expression.

WHAT we will need:
Zinc metal (used for window screens)
Construction paper (white and brown)
People pattern (provided on page 190; photocopy and enlarge for your use)
Tub of water
Popsicle stick
Eyedropper
Small container of liquid soap
Markers
Crayons
Children's scissors

HOW we will do it: Zinc screening is available at hardware stores. To make a raft, start with a zinc rectangle about 12.5 x 7.5 cm (5" x 3") and fold all sides so that the edges are turned up like a box lid. Your corners will poke out, which is fine, although make sure none of the corners are jagged. Put the raft, a tub of shallow water, the container of liquid soap, eyedropper, and Popsicle stick on one activity table. Use the pat-

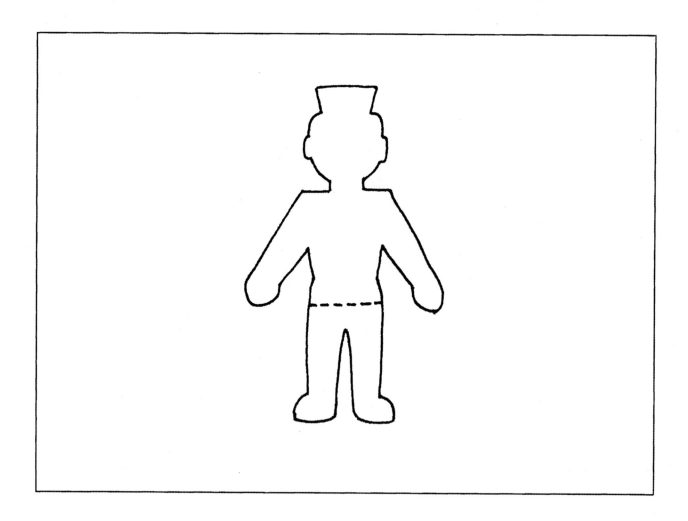

tern in the text to cut out little people from brown and white paper. Put these on another activity table nearby, along with extra brown and white paper, crayons, markers, and children's scissors.

During an Attention Getter time, pass the raft around so that the children can examine it. Even though it has holes, do they think it will float? Encourage predictions. Show your children how to place the raft gently on the surface of the water. What stops it from sinking? (The water's surface tension.) Gently place a few paper people in the raft, and show the children the Popsicle stick, eyedropper, and liquid soap. Ask them to predict what will happen to the raft when soap is dropped on it, but first, show them the rest of the paper people with the art materials, and encourage them to make their own paper sailors. Have water refills handy, as well as a large empty tub for dumping used soapy water. Make several rafts if possible, but

if you only have one raft, make a sign limiting the number of children who may explore the materials at any one time. Encourage the children to verbalize their findings as they conduct this experiment.

Developmental differences: Three- and young four-year-olds may be more interested in engaging in dramatic play with the sailors and boats. If you like, conduct the experiment yourself and see if any children become interested. Older children are likely to be curious about how the soap affects the surface tension and, therefore, the boats.

Soapy Spiral
Science

WHY we are doing this project: to provide children with another fun way of exploring surface tension.

WHAT we will need:
Spiral patterns (provided below; photocopy and enlarge for your use)
Cardboard milk cartons
Water
Liquid soap
Eyedroppers
Small containers
Tubs of water

HOW we will do it: To prepare, use the spiral pattern in the text to cut out spirals from milk cartons. Set them out beside the tubs of water. Put liquid soap and eyedroppers in the small containers.

Ask your students to predict what will happen if they drop the spirals in the water, and use the eyedroppers to drop soap into the spiral centers. Invite your students to conduct this experiment. What happens? (When the surface tension inside the spiral is weakened, the stronger surface tension outside the spiral makes it spin.)

Stream Squeeze
Science/Sensory

WHY we are doing this project: to provide children with a fun, sensory-oriented way of exploring water surface tension and soap's effect on it.

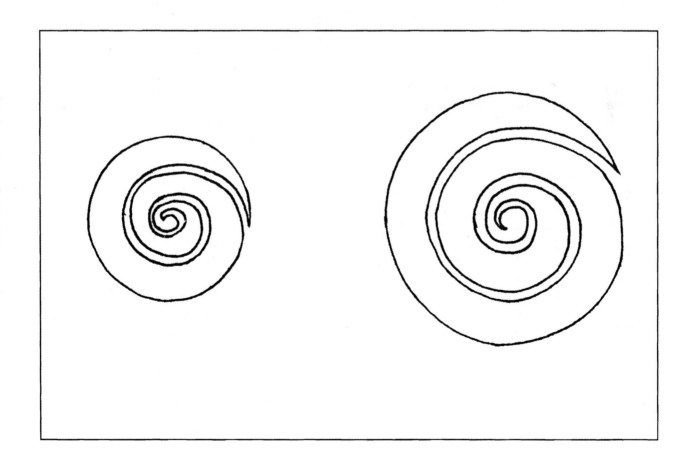

WHAT we will need:
  Small coffee cans
  Clean water
  Small container of liquid soap
  Eyedropper
  Tubs or sensory table
  Plastic bowls with flat bottoms
  Small pitchers
Preparation:
  Hammer
  Sharp nail

HOW we will do it: To prepare, use the hammer and nail to punch three holes, close together in a row, about half an inch from the bottom of each coffee can. In the empty tub or sensory table, turn the bowls upside down and put the tins on top.

Invite the children to cover the holes with their fingers while they fill the tins with pitchers of water. Ask the children to predict, when they take their fingers away, whether the water will run out of all the holes. As they conduct the experiment, ask them to try to squeeze the three streams into one. Why are they able to do this? (The surface tension of all three streams combines to keep them together as one.) After the children have explored this thoroughly, put the liquid soap and eyedroppers out. Ask them to predict whether or not this experiment will work when soap is mixed with the water. Encourage the children to verbalize their findings, and to hypothesize about why the experiment does not work with liquid soap in the water.

Developmental differences: Three- and young four-year-olds will have fun pouring, splashing, and feeling the water, as well as filling containers with the water that streams out of the coffee can. Provide small containers for this kind of sensory exploration. If you like, squeeze the streams together yourself, and see if any children become interested in this experiment. Older children tend to be more fascinated with the phenomenon of squeezing the three streams into one, and with what happens when soap is added.

The Exploding Triangle Trick
Science

WHY we are doing this experiment: to provide children with a dramatic way of observing how soap weakens surface tension; to develop self-esteem and a sense of autonomy through use of a one-person work station.

WHAT we will need:
  Shallow tub of water
  Water refills
  Tub for emptying used water
  Three 10 cm (4") straws
  One Popsicle stick
  Small container of liquid soap
  "One person may be here" sign
    (provided on page 165; photocopy and
    enlarge for your use)
  Newspaper

HOW we will do it: To prepare, spread several layers of newspaper underneath your activity area. Pour a shallow amount of water in the tub, set all other materials beside it, and pin up your "One person may be here" sign.

During an Attention Getter time, read or interpret the sign together. Show the children how to use the Popsicle stick to arrange the straws into a triangle in the water. What will happen if they dip a Popsicle stick into the liquid soap, and then dip it into the middle of the triangle? Encourage predictions and then invite them to find out. Explain that every person who conducts the experiment will have to have clean water. Show them where the refills are, and the empty tub in which used water can be dumped.

Ask your students to hypothesize about why the triangle explodes. When the surface tension in the middle is broken up, the stronger surface tension on the outside pulls the straws apart. After a while, provide an eyedropper. Experiment with dropping different amounts of liquid soap into the middle.

Developmental differences: Three- and young four-year-olds will enjoy splashing and stirring soap into the water with the straws and Popsicle sticks. If you like, conduct the exploding triangle experiment yourself and see if any children become interested. Older children are likely to be more interested in how the straws "explode" when soap is added to the water.

Soap and Light: Making Rainbow Bubbles

Science

WHY we are doing this project: to enable children to observe how light reacts to bubble film.

The facts of the matter: As the walls of a bubble become thinner, white light rays no longer pass through the bubble. Instead, the light is reflected back, either from inside the soapy film or from the outside. When this happens, the colors of the spectrum appear. The colors will change by disappearing and reappearing as the bubble constantly stretches into thinner and thicker film. This explanation may be too involved for young children, but they will enjoy seeing the colors of the spectrum as the bubble film reflects the white light.

WHAT we will need:
Bubble liquid:
6 cups water
2 cups clear dishwashing liquid
   (Joy is recommended)
1 to 4 cups glycerine
Metal coat hangers
Bubble blowers
Bubble holders
Shallow containers
Pliers
Flashlights

HOW we will do it: Mix together the ingredients for the bubble liquid and pour it into the shallow containers. Unbend the coat hangers, and use the pliers, if necessary, to reshape them into bubble blowers. Spread several layers of newspaper under your work area. Show your students the flashlights and invite them to pre-

dict what they will see if they shine light into bubble film. Invite them to use the materials to make bubbles and to conduct this experiment. Mention to the children that they may need to ask each other for help; it might be difficult to hold a bubble blower and a flashlight at the same time. Encourage your students to verbalize their findings.

Super Bubbles

Science/Gross Motor

WHY we are doing this experiment: to enable children to discover that soap makes water elastic; to develop the large muscle group.

The facts of the matter: Soap forces water molecules apart, and this makes water more elastic and "stretchable."

WHAT we will need:
Bubble mixture (several batches)
Wire coat hangers
Hula hoops (if possible)
Straws
String
Large tubs
Trays
Windy day
Preparation:
Pliers

HOW we will do it: Mix up your bubble liquid and put it in the large tub and trays. Untwist the wire coat hangers, and twist them around a cooking pot into large, circular bubble blowers. You will probably need pliers to help you with this. Be sure to leave a handle sticking out from the circle. If you have any hula hoops, these work well as bubble blowers too. To make rectangular super bubble blowers, feed long lengths of string through four straws, and tie the ends together. Slide the straws so that each side is even, and the bubble blower can be held in a rectangular shape.

Put all the materials outside, preferably on a windy day, and encourage the children to use them. Explain that soap makes water molecules move apart, and that this makes the water more "stretchable." Thus, it is possible to blow giant bubbles.

Bubble Blow-Up
Science

WHY we are doing this project: to enable children to experiment further with the elasticity of bubble film; to enable children to discover that the wall of a bubble is so elastic that it will stretch around a straw opening without bursting; to facilitate cooperation between children.

WHAT we will need:
    Bubble wands (regular size)
    Bubble solution
    Straws
    Small containers
    Paper towels

HOW we will do it: To prepare, put some bubble solution in the small containers, and set out the materials on the activity table so that all children will have access to them. During an Attention Getter time, make a bubble on a bubble wand so that the bubble hangs upside down. Ask a child to hold the wand for you in that position. Take a straw and dip it into the bubble solution, making sure the children notice this step. (A dry straw will burst the bubble.) Ask the children to predict whether, when you gently and slowly insert the straw into the upside down bubble, the bubble will burst. Ask them to predict what will happen if you put the straw in the bubble and blow. Ask the children if they see anything in the room that would help them conduct these two experiments, and invite them to do so.

If you work with three- and young four-year-olds, practice blowing out through the straws instead of sucking in. Before the children approach the table, caution them not to

share their straws. Also, say, "I had to ask someone to help me do this experiment, because I couldn't hold the wand very easily while I put the straw into the bubble. When you do this experiment, you may have to ask a friend to help you, and then maybe you can help your friend in the same way."

Depending on your goals for this activity, another option is to spread soapy water on a table surface, and let the children experiment there. This method doesn't require cooperation between students; however, by using this method, we have managed to inflate bubbles to amazing sizes.

Encourage the children to verbalize their findings as they experiment. Ask them if they remember what they learned in the previous activity about how soap makes water more "stretchable." Using this information, ask your students to hypothesize why a bubble does not burst when a wet straw is put inside it. (The bubble wall stretches and bends to let the straw in.) Ask them to hypothesize why the bubble gets bigger when air is blown into it, instead of bursting right away.

More Bubble Tricks
Science

WHY we are doing this activity: to allow children to discover that soap keeps water (and therefore bubbles) from drying out; to provide children with more fun ways of observing that soap makes water elastic.

WHAT we will need:
    Small containers
    Bubble solution
    Bubble wands
    Straws
    Paper towels

HOW we will do it: Set out all materials, and encourage the children to try these experiments:

Catching bubbles and observing a floating bubble land: Ask the children to try to catch a bubble with dry hands. What happens? What

194

happens when a bubble lands on a dry surface? Spread bubble liquid onto a table and encourage the children to blow bubbles over its soapy surface. What happens when the bubbles land on it? Ask the children to hypothesize about the reason for the difference in reactions when a bubble lands on a dry and wet surface. (A dry surface makes a bubble pop right away, but soap keeps the bubble from drying out, so it lasts longer.) Invite the children to put bubble liquid on their hands and to predict whether a bubble will burst if they try to catch it. Invite them to conduct this experiment. (Sometimes the impact of the landing bubble will cause it to burst, even on a soapy hand, but other times the bubble will land on a soapy hand and stay intact for several seconds.)

Blowing a bubble within a bubble: Slick down a table surface with bubble liquid. Encourage the children to blow bubbles so that they land on the soapy surface. Invite them to dip a straw into bubble liquid, insert the straw into a large bubble, and blow a new, smaller bubble inside. If you provide several containers of bubble liquid, each with a different food coloring mixed in it, this experiment can be especially interesting.

Bubble squeeze: Show the children how to hold a bubble wand in each hand, and gently blow a bubble through one of the wands (without blowing it off). Ask them to predict what will happen if they dip the empty wand into bubble liquid, put it on the other side of the bubble, and squeeze. Encourage your students to verbalize their observations.

What happens when they gently and slowly pull the wands away from each other? (The bubbles stretch and change shape.) What happens if they slide a soapy wand over a bubble that is already sitting on another bubble wand?

Grating Soap
Gross Motor/Sensory

WHY we are doing this activity: to provide a sensory experience for children and to develop the large muscle group.

WHAT we will need:
    Soap bars
    Freestanding graters
    Tubs of water
    Paper towels
    Play dishes

HOW we will do it: Set all materials on an activity table. Invite the children to grate soap and use it in water to clean the play dishes, or wash plastic toys. What does the grated soap feel like, compared to bar soap? What happens to them both in water? What do the children notice about the difference in the kinds of bubbles they make?

Soap Boxes: Math Kits
Math

WHY we are doing this project: to practice rational counting and, for older preschoolers and kindergartners, subtraction.

WHAT we will need:
    Small, shaped soaps
    Three boxes with lids to hold collections
        of soaps
    Blank sheets of paper
    Writing sheets (format provided on
        page 196; photocopy and enlarge
        for your use)
    Small pencils

HOW we will do it: Many gift shops have a large variety of small, shaped soaps that cost about $.35 to $.85. Buy enough of these to put a

How many soaps are in the 🧴? _____

Take three soaps away. How many are left? _____

How many soaps are in the 🧴? _____

Take _____ soaps away. How many are left? _____

Use this sample to draw pictures that represent the box and soaps you are actually using, as well as the number of soaps you want taken away. Include the subtraction problem only if you are working with older preschoolers or kindergartners.

small collection in each box. Make several copies of the writing sheet, and put one of these (or blank sheets of paper that are the same size) with a pencil in each box. Keep spares handy so that you can replace used sheets with blank ones as the children use the materials.

Set the soap boxes out on the floor or a table. During an Attention Getter time, show one to the children and ask them to guess what is inside. Take out the writing sheet, and together, read or interpret what it says. As a group, count the number of soaps in the box, and on the blank piece of paper or writing sheet, write down the number you counted. With kindergartners, read/interpret the subtraction problem and have the children help you count the number of soaps to take away. Count the soaps that are left, and write this number on the blank paper or writing sheet. After your demonstration, change the number of soaps in that box so that the math problem will be a new one. Invite the children to explore the soap boxes, and replace paper as necessary.

Soap Match
Math

WHY we are doing this activity: to facilitate a matching exercise; to provide a math experience for younger children; to promote self-esteem through use of a one-person work station. (For younger preschoolers.)

WHAT we will need:
Collection of shaped soaps
Butcher paper
Dark-colored marker
Basket
"One person may be here" sign
(provided on page 165; photocopy and enlarge for your use)

HOW we will do it: To prepare, arrange your soaps on the paper. Trace around them with the marker. If you have more than one soap of the same shape, make sure they are different colors and shade the corresponding outline in the corresponding color. Put the soaps in a basket, next to the butcher paper. If you are going to be using this activity again, draw your outlines on poster board and cover with contact paper.

During an Attention Getter time, show the children the materials you will be using, and discuss the "One person may be here" sign. Invite your students to match the soaps to their outlines.

Soap Book
Science/Language/Art/Fine Motor

WHY we are doing this project: to reinforce the fact that soap comes in many different forms; to develop the small muscle group; to facilitate creative expression; to develop all components of language arts: reading, writing, speaking, and listening.

WHAT we will need:
    Construction paper
    Family/home magazines
    Soap and bubble pictures (provided
        below; photocopy and enlarge for
        your use)
    Scissors
    Glue
    Small containers for glue
    Markers
    Crayons
    Pens
Preparation:
    Stapler

HOW we will do it: Begin collecting family/home magazines several weeks ahead of time. Ask friends, family, and other parents to help. To make one blank soap book, put two pieces of construction paper on top of each other, and fold into equal halves. Staple along the fold.

Color in the provided soap and bubble pictures. Affix these to the cover and print: "The Soap Book." Make one of these books for each child. Leave several covers blank for children who want to design their own.

Leaf through the magazines, and tear out photographs of bar soap, liquid soap, dishwashing liquid, and powdered detergent. Ahead of time, make a sample soap book by gluing some of these pictures onto the pages of your book and writing words. (Example: "This is the kind of soap I use to wash my clothes. It feels gritty when I touch it." "This bar of soap is a pretty color. Lilacs are flowers that are the same color.")

Set out all materials on the table so that all children will have access to them. During an Attention Getter time, show the children your sample soap book and read it to them.

Ask your students if they see anything in the room that will help them make their own soap books. If you work with young preschoolers, take story dictation. If you work with older children, spell words, support invented spelling, or write words down on separate pieces of paper to be copied, according to what the children request. When the books are completed, ask the children to read them back to you, tell you about them, or read the books to them. Ask if you can put them out on your bookshelf with your other books.

Bubble Art
Science/Art/Sensory

WHY we are doing this activity: to enable children to see that soap, force, and air make millions of tiny bubbles; to allow children to express themselves artistically with different materials; to provide them with a sensory experience.

WHAT we will need:
    1 cup Ivory Snow flakes
    1 cup water
    Food coloring (variety of colors)

Small containers
Paper
Large bowl
Newspaper
Magnifying glass
Preparation:
    Electric beater

HOW we will do it: Spread several layers of newspaper under your work area. Use the electric beater to whip the water and soap flakes until the mixture is thick. If you like, have the children help you do this. Put some of the mixture into the small containers and to each one add some food coloring. Encourage the children to examine the mixture with the magnifying glass. What do they see? Ask them how many bubbles they think there might be in the mixture. Encourage the children to finger paint with the soapy mixture. When the pictures dry, what do the children see (tiny bubble prints from dried bubbles)?

Bubble Prints
Science/Art

WHY we are doing this project: to enable children to see that burst bubbles leave prints behind; to facilitate creative expression; to facilitate cooperation through a group project. (This project makes gorgeous paper!)

WHAT we will need:
    Bubble mixture
    Food coloring
    Bubble blowers/wands
    Butcher paper (white)
    Small containers
Preparation:
    Masking tape

HOW we will do it: To prepare, use masking tape to secure butcher paper over your activity table, so that the entire surface is covered. Put bubble mixture into each small container and add a different color of food coloring to each one. Use various bubble blowers and wands, for example, fly swatters with lattice-style plastic, twisted wire, and plastic six-pack bottle holders.

Set all the materials out in the middle of the table, on top of the butcher paper, and encourage the children to blow bubbles over it, so that they burst on the paper. As the children do this, watch the paper get prettier and prettier. If you like, provide certain combinations of colors, such as: brown/orange/yellow, purple/pink/turquoise, or green/blue/aquamarine. Make turquoise by adding a little bit of green to blue. Make aquamarine by adding a little bit of blue to green. When the bubble-print picture is finished, hang it up on the wall, or cut it into pieces and use it as gift wrap for presents the children make.

Painting Soap
Art

WHY we are doing this project: to facilitate creative expression and to develop fine motor skills.

WHAT we will need:
> Bars of soap
> Water-based paints
> Fine paintbrushes
> Newspaper
> Small containers of water
> Small tub of water

HOW we will do it: Spread newspaper out on your work surface, and put all materials on the table. Invite the children to paint the soaps, and sit down and paint one yourself. Show the children how to wash the soaps in the small tub of water when they want to paint a new design on the same bar of soap.

Laundry Day
Dramatic Play/Language/Gross Motor/
Multicultural/Anti-Bias

WHY we are doing this activity: to give children the opportunity to work through emotions during pretend play; to facilitate social interaction between children; to help children develop conversational skills and the ability to put their ideas into words; to help children develop awareness of diverse cultures and gender roles.

WHAT we will need:
> Large tubs or sensory table
> Ivory Snow flakes
> Miniature boxes of soap detergent
> Dolls' clothes and dress-up clothes
> Clothes rack
> Clothes wringer (if possible)
> Clothespins
> Newspaper
> Magazines (including *National Geographic*)

Clean-up:
> Mop

HOW we will do it: Spread many layers of newspaper under your "laundry" area. You can buy miniature boxes of laundry detergent at variety stores, sometimes as promotional samples, or in vending machines at laundromats. Replace the detergent in them with Ivory Snow flakes. Fill several of your tubs with cold water for rinsing. Put all of the clothes in another tub. If you can find an old-fashioned laundry wringer, children love using these.

Leaf through your magazines, including *National Geographic*, for photographs of people washing clothes, and tape these on the walls around the laundry area. Aim for a collection of photos that is culturally diverse and, if possible, in opposition to gender stereotypes. Invite your students to explore the materials. Children absolutely love this dramatic play, but it does tend to get messy, so be prepared with your mop when it is all over.

Literature

Symbol key: *Multicultural
             +Minimal diversity
             No symbol: no diversity or no people

Bell, J. L. (1993). *Soap science.* Redding, MA: Addison-Wesley.

Sutherland, H. A. (1988). *Dad's car wash.* New York: Atheneum.

The following books feature bubbles:

   Woodruff, E. (1991). *Show and tell.* New York: Holiday House.

   Woodruff, E. (1990). *Tubtime.* New York: Holiday House.

Extenders

Music: Here is a good song to sing when you make the super bubbles outside.

To the tune of "Oh my darling Clementine":
>   Super bubble, super bubble,
>   watch me blow it in the air,
>   wind is blowing, bubble's floating,
>   whoops it burst, right over there!

Language: If you choose to hang the bubble-print paper on the wall, ask the children later if they remember how they made it, and write their comments on the top, bottom, and sides of the paper. If you cut the paper into gift wrap, make gift tags too. Take dictation for the gift tags, or if you work with older children, have them write their own words. You could give shaped soaps or the painted soaps as gifts.

Science: This is an extender for the whipped-soap painting. Use an electric mixer, soap, and cold and warm water to see if the same amount and kinds of bubbles are produced. Does the force created by the electric mixer make a difference?

Dramatic Play: Put containers for various types of soaps (e.g., dishwashing liquid, Ivory Snow flakes, soap bars, and soap leaves) in your housekeeping area or in a dramatic-play store.

INDEX

Other Titles Published by Delmar Publishers

The Essentials of Early Education

Carol Gestwicki, M.A.

Central Piedmont Community College, North Carolina

Young children need people with specific knowledge, skills, and attitudes, who have thoughtfully prepared to enter into caring relationships with young children and their families. This book examines the world of early education and assists in the process of professional growth for those who are interested in considering it as their future. Designed to be the ideal motivational tool for beginning teachers entering the field, The Essentials of Early Education encourages students to be active participants in the decision making process of becoming early childhood teachers. All the phases of early childhood education are thoroughly covered: the scope of children served, the types of programs, different styles and philosophies of teaching, in addition to defining all the aspects of a quality education and the teacher's role in education today.

FEATURES:

- Written in brief, interactive style that is career focused.
- Full color art program.
- "Theory into Practice" section in each chapter introduces real teachers and issues.
- Supported with multiple classroom activities and quizzes to enhance analytical and critical thinking skills.
- Unique "Time Line of ECE" feature runs as a footnote throughout the text with an event or person and its relevance to early childhood or development.
- A running glossary appears in the column so that students have all the information they need to know at their fingertips.

TABLE OF CONTENTS:

THE STARTING POINT. Decision Making. Early Childhood Education Today. What Quality Early Education Looks Like. What Teachers Do. DEVELOPING AS A TEACHER. Why Become a Teacher? Growing Oneself as a Teacher. Challenges for Early Educators. THE PROFESSION COMES OF AGE. Roots of Early Education. The Modern Profession. Professional Education & Career Directions. Current Issues in Early Education. The Road Ahead.

355 pp., softcover, 8" x 10"
Text 0-8273-7282-5
Instructor's Manual 0-8273-7283-3

To Order Call Toll Free
1-800-354-9706
Fax: 1-800-487-8488
or use the Order Form on the
last page of this book.

Health, Safety & Nutrition for the Young Child, 4E

Lynn R. Marotz, Ph.D., University of Kansas
Marie Z. Cross, Ph.D., University of Kansas
Jeanettia M. Rush, R.D., M.A.

We are pleased to present the fourth edition of the best selling textbook in early education. This full color book sets the standard for the three most crucial areas of child development: children's health status, a safe, yet challenging learning environment and proper nutrition. The new edition includes updated information on the most current issues in child care. Emphasis is given to the topic of quality child care and organizing quality care environments for children. This edition includes increased coverage of AIDS and children, Attention Deficit Disorder (ADD), ADHD, Sudden Infant Death Syndrome (SIDS), lead poisoning, diabetes, seizures, allergies, asthma, eczema, sickle cell anemia, immunization, emergency care, and common illnesses as well as life threatening conditions.

FEATURES:
- Full color illustrations and photographs bring theory and practice alive!
- Key words are noted in color, italicized, and defined in the glossary (reinforcing the students' analytical and critical thinking skills).
- Each chapter contain summaries, hands-on learning activities and review questions to foster active learning skills
- Electronic study guide, packaged with text, includes answers to review questions, test items, resources and discussion topics.

WHAT'S NEW
- The new "Food Guide Pyramid"
- Toddler feeding
- Optional electronic study guide
- Infant feeding concerns
- Sample activity plans

TABLE OF CONTENTS:
HEALTH, SAFETY, & NUTRITION: AN INTRODUCTION: Interrelationship of Health, Safety, & Nutrition. HEALTH OF THE YOUNG CHILD: MAXIMIZING THE CHILD'S POTENTIAL: Promoting Good Health. Health Appraisals. Health Assessment Tools. Conditions Affecting Children's Health. The Infectious Process & Effective Control. Communicable & Acute Illness: Identification & Management, SAFETY FOR THE YOUNG CHILD: Creating a Safe Environment. Safety Management. Management of Accidents & Injuries. Child Abuse & Neglect. Educational Experience for Young Children. FOODS & NUTRIENTS: BASIC CONCEPTS: Nutritional Guidelines. Nutrients That Provide Energy. Nutrients That Promote Growth of Body Tissues. Nutrients That Regulate Body Functions. NUTRITION & THE YOUNG CHILD. Infant Feeding. Feeding the Toddler & the Preschool Child. Planning & Serving Nutritious Meals. Food Safety & Economy. Nutrition Education Concepts & Activities. APPENDICES: Nutrition Analysis of Various Fast Foods. Growth Charts for Boys & Girls. Sources of Free & Inexpensive Materials Related to Health, Safety, & Nutrition. Federal Food Program. Glossary. Index.

506 pp., softcover, 7 3/8" x 9 1/4"
Text 0-8273-8353-3   Instructor's Guide 0-8273-7274-4

Flash!

Seeing Young Children: A Guide to Observing and Recording Behavior, 3E

Warren R. Bentzen, Ph.D.
State University of New York at Plattsburgh

Seeing Young Children provides the essential guidelines for observing young children. This text includes the steps one must take before entering the observation setting to professional ethics and confidentiality.

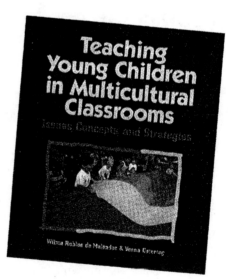

TABLE OF CONTENTS:
OVERVIEW: Introduction. Overview of Developmental Theories. General Guidelines for Observing Children. THE ELEMENTS OF OBSERVATION: Methods, Behaviors, Plans, and Contexts. An Introduction to Observation & Recording Methods. Narrative Descriptions. Time Sampling. Event Sampling. Diary Description. Anecdotal Records. Frequency Counts or Duration Records. Checklists. Application: Recording Methods in Action. Interpretations of Observations, Implementation of Findings, and Ongoing Evaluation. OBSERVATION EXERCISES: Introduction & Preparation. Observing the Newborn: Birth to One Month. Observing the Infant (One to Twenty-Four Months). The Young Child: Ages Two Through Five. MIDDLE CHILDHOOD: The School-Age Years. Introduction & Preparation. The School-Age Years: The Six-Year Old Child. The School-Age Years: The Seven- and Eight-Year Old Child.

384pp., softcover, 7 3/8" x 9 1/4"
Text 0-8273-7665-0
Instructor's Manual 0-8273-7666-9

Teaching Young Children in Multicultural Classrooms

Wilma Robles DeMelendez, Ph.D.
Nova Southeastern University, Florida
Vesna Ostertag, Ed.D.
Nova Southeastern University, Florida

Teaching Young Children in Multicultural Classrooms is a comprehensive study of the historical, theoretical and practical aspects of multicultural education as it relates to young children. This book includes comprehensive current and future trends, and provides many practical classroom ideas.

TABLE OF CONTENTS:
Facing the Reality of Diversity: The Intricate Nature of Our Society. The Nature of Culture, The Nature of People. Families in Our Classrooms: Many Ways, Many Voices. Who is the Child? Developmental Characteristics of Young Children. Everything Started When.....Tracing the Beginnings of Multicultural Education. Approaches to Multicultural Education: Ways & Designs for Classroom Implementation. The Classroom, Where Words Become Action. Preparing to Bring Ideas into Action. Activities & Resources for Multicultural Teaching: A World of Possibilities! A World of Resources: Involving Parents, Friends, & the Community. Appendix.

416pp., softcover, 8" x 9 1/4"
Text 0-8273-7275-2
Instructor's Manual 0-8273-7276-0

Early Childhood Curriculum: A Child's Connection to the World

Hilda Jackman, Professor Emerita
Brookhaven College, Dallas County Community College, Texas

This innovative text presents developmentally appropriate early education curriculum in a clear easy-to-read style. All chapters of the text stand alone, while complementing each other to form the whole curriculum for children from infancy to eight years. The text includes original songs, poems, dramatic play activities, as well as numerous illustrations, photos, diagrams, references, and teachers' resources.

TABLE OF CONTENTS:
PART 1: Creating the Environment That Supports Curriculum & Connects Children: STARTING THE PROCESS: CREATING CURRICULUM: PART 2: Discovering & Expanding the Early Education Curriculum: LANGUAGE & LITERACY: LITERATURE: PUPPETS: DRAMATIC PLAY & CREATIVE DRAMATICS: ARTS: SENSORY CENTERS: MUSIC & MOVEMENT: MATH: SCIENCE: SOCIAL STUDIES: APPENDICES: GLOSSARY: INDEX.

368 pp., softcover, 8 1/2" x 11"
Text 0-8273-7327-9
Instructor's Guide 0-8273-7328-7

Growing Artists: Teaching Art to Young Children

Joan Bouza Koster
Broome Community College, New York

Growing Artists will prepare pre-service teachers to teach art to children ranging from one and a half through eight years old. Each chapter focuses on a particular topic in art education, including current theory and research, the role of the teacher, how children develop artistically, creating an aesthetic environment, and integrating art into the curriculum.

TABLE OF CONTENTS:
Growing Young Artists. The Artist Inside. How Young Artists Grow. A Place For Art. Making Connections. Please Don't Eat The Art! I'm Creative. Anyone Can Walk On The Ceiling. It's A Mola. Growing Together! Caring & Sharing. Have We Grown? Appendices. References. Index.

448 pp., softcover, 8 1/2" x 11"
Text 0-8273-7544-1
Instructor's Manual 0-8273-7545-X

Student Teaching: Early Childhood Practicum Guide, 3E

Jeanne M. Machado, Emerita, San Jose City College, California
Helen Meyer-Botnarescue, Ph.D.
California State University at Hayward

Student Teaching covers all aspects of teaching from understanding children and staff communication to relationships with parents. This book includes the characteristics of contemporary American families through multicultural and ethnic diversity. Contains chapters on dealing with children with special needs, infant and toddler placement, professionalism, trends, and issues in education.

TABLE OF CONTENTS:

464 pp., softcover , 8″ x 9 1/4″
Text 0-8273-7619-7   Instructor's Guide 0-8273-7620-0

Building Understanding Together: A Constructivist Approach to Early Childhood Education

Sandra Waite-Stupiansky, Ph.D.
Edinboro University of Pennsylvania

Based on sound learning, *Building Understanding Together: A Constructivist Approach to Early Childhood Education* demonstrates the basic tenets of Piaget's constructivist theory in a comprehensive format. This text shows how constructivism can be applied to all areas of the curriculum; language arts, science, math, social studies, and the arts.

TABLE OF CONTENTS:

210 pp., softcover, 6″ x 9″
Text 0-8273-6835-6   Instructor's Guide 0-8273-6836-4

Week by Week: Plans for Observing and Recording Young Children

Barbara Ann Nilsen, Ed.D.
Broome Community College, New York

This well-organized book provides students with a systematic plan for week-by-week documentation of each child's development in an early childhood setting. It presents instruction in the most common recording techniques as well as a review of basic child development principles. By following the week-by-week plan, the observer is able to collect periodic and useful information from a variety of sources. Diagrams, thinking exercises, and case-studies help the content meet the individual learning styles of the readers and the weekly assignments provide manageable collection of data to form a portfolio illustrating the child's development in all areas.

TABLE OF CONTENTS:

416 pp., softcover, 8 1/2" x 11"
Text 0-8273-7646-4
Instructor's Guide 0-8273-7647-2

Other Favorites Available from Delmar!

474 Science Activities for Young Children,
Green, 0-8273-6663-9

Administration of School for Young Children,
Click, 0-8273-5876-8

*Art and Creative Development for Young
Children, 2E,* Schirrmacher, 0-8273-5776-1

Assessing Young Children, Mindes, Ireton, and
Mardell-Czudnowski, 0-8273-6211-0

*Beginnings and Beyond: Foundations in Early
Childhood Education, 4E,*
Gordon & Browne, 0-8273-7271-X

Creative Activities for Young Children, 5E,
Mayesky, 0-8273-5886-5

*Creative Resources for the Early Childhood
Classroom, 2E,* Herr & Libby, 0-8273-5871-7

*Developing and Administering a Child Care
Center, 3E,* Sciarra & Dorsey, 0-8273-5875-4

*The Developmentally Appropriate Inclusive
Classroom in Early Education,*
Miller, 0-8273-6704-X

Developmentally Appropriate Practice,
Gestwicki, 0-8273-7218-3

Developmental Profiles: Prebirth to Eight, 2E,
Allen & Marotz, 0-8273-6321-4

*Early Childhood Curriculum: From
Developmental Model to Application,*
Essa & Rogers, 0-8273-7483-6

*Early Childhood Experience in Language Arts:
Emerging Literacy, 5E,*
Machado, 0-8273-5242-5

*Emergent Literacy and Dramatic Play in
Early Education,* Davidson, 0-8273-5721-4

*The Exceptional Child. Inclusion in Early
Childhood Education, 3E,*
Allen, 0-8273-6698-1

Experiences in Math for Young Children, 3E,
Charlesworth, 0-8273-7226-4

*Experiences in Movement with Music,
Activities and Theory,* Pica, 0-8273-6478-4

*Exploring Science in Early Childhood: A
Developmental Approach, 2E,*
Lind, 0-8273-7309-0

Growing Up with Literature, 2E,
Sawyer & Comer, 0-8273-7228-0

A Guidance Approach to Discipline,
Gartrell, 0-8273-5520-3

*Home, School and Community Relations: A
Guide to Working with Parents, 3E,*
Gestwicki, 0-8273-7218-3

Infant and Child Care Skills,
Bassett, 0-8273-5507-6

*Infants and Toddlers: Curriculum and
Teaching, 3E,* Wilson, Watson & Watson,
0-8273-6094-0

*Integrated Language Arts for Emerging
Literacy,* Sawyer & Sawyer, 0-8273-4609-3

Introduction to Early Childhood Education,
2E, Essa, 0-8273-7483-6

Math and Science for Young Children, 2E,
Charlesworth & Lind, 0-8273-5869-5

The Montessori Controversy,
Chattin-McNichols, 0-8273-4517-8

Positive Child Guidance, 2E, Miller,
0-8273-5878-4

*Practical Guide to Solving Preschool
Behavior Problems, 3E,* Essa, 0-8273-5812-1

Science Is Fun!, Oppenheim, 0-8273-7336-8

*Stories: Children's Literature in Early
Education,* Raines & Isbell, 0-8273-5509-2

Topical Child Development, Berns,
0-8273-5727-3

Understanding Child Development, 4E,
Charlesworth, 0-8273-7332-5

To Order Call Toll Free
1-800-354-9706
Fax: 1-800-487-8488
or use the Order Form on the
last page of this book.

To Place Your Order

Simply fill-in the following information and mail this form to:

Attn: Order Fulfillment
7625 Empire Dr.
Florence, KY 41042

Or, call **1-800-354-9706**

ORDER #	AUTHOR/TITLE	QUANTITY	PRICE	TOTAL
08273-6663-9	Green/474 Science Activities			
07668-0010-5	Green/Not! The Same Old Activities for Early Childhood			
07668-0009-1	Green/Themes with a Difference: 228 New Activities for Young Children			
08273-5871-7	Herr/Creative Resources for the Early Childhood Classroom, 2e			
07668-0015-6	Herr/Creative Resources of Art, Brushes, Buildings . . .			
07668-0016-4	Herr/Creative Resources of Birds, Animals, Seasons, and Holidays			
07668-0017-2	Herr/Creative Resources of Colors, Food, Plants, and Occupation			
08273-7281-7	Wheeler/Creative Resources for the Elementary Classroom and School-Age Programs			
08273-8094-1	The Best of Wonderscience: Over 400 Hands-On Elementary Science Activities			
08273-7107-1	Wynn/Creative Teaching Strategies: A Resource Book for K-8			
			Subtotal*	
			Add State, Local Taxes	
			TOTAL ORDER	

Charge to your credit card or enclose payment with your order and Delmar will
pay postage and handling. Your local sales tax must be included with payment.

SELECT PAYMENT METHOD

❏ Enclosed is my purchase order number _____

❏ A check is enclosed for $_____ (including sales tax)

❏ Charge my ❏ *VISA* ❏ MasterCard Card # _____ Exp. Date _____

Signature _____

Name _____ Position/Title _____

School/Institution _____ Phone No. (Office) _____

School Address _____ City _____ State _____ Zip _____

❏ Please have a Delmar representative contact me.

PRICES SUBJECT TO CHANGE WITHOUT NOTICE.
APPROPRIATE DISCOUNTS WILL BE APPLIED.